Cambridge astrophysics series

T0171986

The physics of solar flares

THE PHYSICS OF
SOLAR FLARES

EINAR TANDBERG-HANSSEN
Marshall Space Flight Center

and

A. GORDON EMSLIE
The University of Alabama in Huntsville

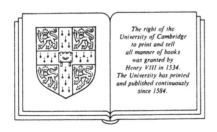

The right of the
University of Cambridge
to print and sell
all manner of books
was granted by
Henry VIII in 1534.
The University has printed
and published continuously
since 1584.

CAMBRIDGE UNIVERSITY PRESS

Cambridge

New York   New Rochelle   Melbourne   Sydney

CAMBRIDGE UNIVERSITY PRESS
Cambridge, New York, Melbourne, Madrid, Cape Town, Singapore, São Paulo, Delhi

Cambridge University Press
The Edinburgh Building, Cambridge CB2 8RU, UK

Published in the United States of America by Cambridge University Press, New York

www.cambridge.org
Information on this title: www.cambridge.org/9780521115520

First published 1988
This digitally printed version 2009

A catalogue record for this publication is available from the British Library

Library of Congress Cataloguing in Publication data
Tandberg-Hanssen, E. (Einar), 1921–   The physics of solar flares.
(Cambridge astrophysics series)
Bibliography: p.
Includes index.
1. Solar flares.   I. Emslie, A. Gordon.
II. Title.   III. Series.
QB526.F6T37   1988   523.7′5   88-2580

ISBN 978-0-521-30804-5 hardback
ISBN 978-0-521-11552-0 paperback

Contents

Units, constants, and symbols

Because of its overwhelming use in the solar physics community, we have, a little reluctantly, decided to use cgs units throughout this book. In addition, we yield to another common practice among solar physicists and discuss the wavelength of radiation in terms of angstroms. The reader will therefore, for example, encounter magnetic fields measured in Gauss ($1\,G = 10^{-4}$ tesla), pressure in dyne cm^{-2}, currents in statamps, and wavelengths of spectral lines referred to in angstroms (1 angstrom $= 10^{-8}$ cm). The different symbols used will be defined in the text, since the same symbol (letter) may be used to denote different parameters or constants. However, when no definition is given, the following letters will always refer to the following physical constants or parameters:

Boltzmann's constant	$k = 1.38 \times 10^{-16}$ erg K^{-1}
Planck's constant	$h = 6.626 \times 10^{-27}$ erg s
Gravitational constant	$G = 6.67 \times 10^{-8}$ dyne cm^2 g^{-2}
Velocity of light	$c = 2.998 \times 10^{10}$ cm s^{-1}
Electron rest mass	$m_e = 9.1 \times 10^{-28}$ g
Proton rest mass	$m_p = 1.67 \times 10^{-24}$ g
Elementary charge	$e = 4.80 \times 10^{-10}$ esu
Thermal conductivity coefficient	$K = 1.0 \times 10^{-6}$ erg cm^{-1} s^{-1} K$^{-7/2}$
Temperature	T (K)
Electron number density	n_e (cm^{-3})
Frequency	$\omega = 2\pi v$ (s^{-1})
Magnetic field	B (G)
Coulomb logarithm	$\ln \Lambda$ [equation (4.64)]
Collisional stopping parameter	$C = 2\pi e^4 \ln \Lambda$ [equation (4.69)]
Bremsstrahlung cross-section coefficient	$\kappa_{BH} = \dfrac{8\alpha}{3} r_0{}^2 m_e c^2$ [equation (5.23)]
Electron plasma frequency	$\omega_{pe} = (4\pi n e^2 / m_e)^{1/2}$ [equation (3.75)]

Preface

Few phenomena have stirred the interest of solar scientists more than the explosive manifestation of energy release we call solar flares. A rich literature exists describing this many-faceted phenomenon, and hypotheses and theories abound trying to explain the different stages observed during a flare. In 1976 Švestka's excellent book *Solar Flares* appeared, and in the late 1970s the Skylab data on flares were studied and a workshop on Solar Flares was conducted by NASA, leading to a comprehensive monograph edited by Sturrock (1980). Another workshop on flares was conducted by NASA in 1983–84, focusing on the excellent results from the Solar Maximum Mission, and the workshop proceedings, edited by Kundu and Woodgate (1986), have recently appeared.

With this background we hesitated at first, when approached by Cambridge University Press, to agree to writing another book on solar flares. However, we also realized that the nature of workshop proceedings generally dictates a very different approach than does the philosophy behind the presentation in a text such as the present volume. We will draw heavily on results from the SMM, but other sources, both from space-borne and ground-based instrumentation, will play crucial roles. In addition, we will incorporate the findings of many new theoretical investigations, always trying to put them in the logical physical context. Consequently, we venture to present *The Physics of Solar Flares* in the hope that its main emphasis – on understanding the physical processes – will make our effort worthwhile.

Solar activity manifests itself in many ways and, while the minor manifestations, for example facular and plage brightening, do not require large amounts of energy, the cataclysmic flare phenomenon severely taxes most available energy sources. Ultimately, one is left with only two: the energy stored in nonpotential magnetic fields and the energy available in the vast photospheric and subphotospheric mass–motion fields. Controversy exists as to how one uses these energy sources to explain solar flares. We can consider solar activity as manifestations of the interaction of magnetic fields with the solar plasma in motion, and this book tries to apply this formula to the flare phenomenon.

The book therefore differs from its predecessors in the sense that it is aimed primarily at beginning graduate (or advanced undergraduate) students who are assumed to be acquainted with the basic knowledge of physics required to understand what we have learned about flares but who need an introduction to the context in which this knowledge is to be applied. The book starts from basic physical principles, such as Maxwell's equations and the equations of fluid mechanics, and develops these into the 'language' of solar flare physicists. Readers who have grasped the content of this book should feel qualified to contribute to current research in solar flare physics and to read published papers in the field critically.

We have divided the book in two main parts. After an overview of the observational aspects of flares, the next four chapters present the tools we deem necessary to attack the physics associated with the flare, and then the following four chapters treat the different phases of the flare or, stated differently, the different manifestations of the flare phenomenon.

The mathematical and physical background that provides the tools for a proper and indepth study of flares consists of Spectroscopy (Chapter 2), Magnetohydrodynamics and Plasma Physics (Chapters 3 and 4), and Radiation Processes (Chapter 5). Thus equipped, we then follow the evolution of the flare process from Pre-flare Conditions (Chapter 6) through the Impulsive Phase (Chapter 7) to the Gradual Phase (Chapter 8). The interaction of flares with prominences and the corona is treated in Chapter 9 in the context of mass ejections. In the Epilogue we offer a brief prognosis for solar flare research in the near future.

Huntsville, June 1987   E. Tandberg-Hanssen
                                                          A. G. Emslie

Acknowledgments

The authors wish to thank a great many individuals who have, in one way or another, contributed to the production of this book. In particular, we are indebted to Drs M. E. Machado and T. N. LaRosa for their helpful criticisms of the manuscript, and to Drs M. R. Kundu and R. L. Moore for many stimulating discussions. A.G.E. would like to give special thanks to Dr J. C. Brown and Professor P. A. Sweet for stimulating his interest in the fascinating subject of solar flares and for guiding him through the difficult graduate student phase that this book is aimed at alleviating.

We are thankful to Mary James, Adonna Mitchell, Tauna Moorehead, and Uli Spradling for painstaking work in typing and editing the manuscript.

To Erna and Buffie

An overview of the solar flare phenomenon

Facts are stubborn things

Gil Blas, A. R. Lesage (1668–1747)

The first recorded observation of the flare phenomenon was made independently by two observers, R. C. Carrington and R. Hodgson, on 1 September 1859. While engaged in the routine survey of sunspots on the solar disk, they witnessed an intense brightening of regions in a complex sunspot group; the event lasted only a few minutes. This was an example of a relatively rare event – a large white-light flare – in which the optical continuum is enhanced sufficiently over the background photospheric field to be visible in contrast, in this case with a relatively crude instrument. Most flares are not so conspicuous in visible light; they reserve their strongest enhancements for spectral lines such as Hα, and they also radiate copious amounts of energy in extreme ultraviolet (EUV) and soft X-ray wavebands. With the advent of spacecraft observations from about 1960 onward, our observational data base on solar flares has increased by orders of magnitude. Flares have been observed across the electromagnetic spectrum, from decametric radio emission to γ-rays in excess of 10 MeV. However, it is fair to say that our theoretical understanding has failed to keep pace with this explosive growth in our observational knowledge. Some of the reasons why this is necessarily true will be dealt with as they arise, and we will, where applicable, point to areas now ripe for rapid development.

1.1 History of observations

Until the time of the Orbiting Solar Observatory series of satellites and the Skylab Space Station with its Apollo Telescope Mount cluster of solar instruments, flare observations were almost uniquely carried out photographically in one of a small number of optical spectral lines, especially in the hydrogen Hα line at 6563 Å. An overwhelming amount of data on different flares therefore exists in the form of Hα only, and general conclusions as to the size, shape, lifetime, intensity, etc., of flares have historically been drawn from these data. In particular, flare classifications, pertaining to an assumed 'importance' of the flares, have been devised. Table 1.2 below shows the dual form for importance classification adopted by the International Astronomical Union (IAU) (1966). The classification describes fairly well the low-energy, cool optical flare whose importance correlates well with many terrestrial, flare-induced effects such as certain geomagnetic storms, auroral displays, etc. On the other hand, such classifications do not, nor can they, reveal the basic physical processes responsible for the flare. Research during the last several years has shown that while the optical flare, as portrayed in $H\alpha$, is an impressive display, there is a hot, high-energy part of most flares, and the study of this part reveals much of the underlying physics of flares. Nevertheless, observations of the cool component, both its spectral signatures and its morphology, offer important clues to its physical character, as stated by Smith and Smith (1963) in their book, which for many years served as an important source of information on the 'classical' flare. While the earlier spectral observations of flares mainly utilized hydrogen lines and a few helium lines to derive physical quantities such as temperature and density, they also provided information on a large number of lines from neutral and singly ionized metals. These latter lines provided information on excitation conditions in flares and, furthermore, served to classify flares and related prominences (Waldmeier, 1951; Zirin and Tandberg-Hanssen, 1960; Tandberg-Hanssen, 1963) in systems relating to the low-energy part of flare plasmas.

Particular interest has always been attached to the observation of white-light flares [e.g., Švestka (1966a); McIntosh and Donnelly (1972); Uchida and Hudson (1972)]. Once considered a rare phenomenon, white-light flares have recently been studied in considerable detail and are now thought of as a fairly common occurrence during sunspot maximum years (Neidig and Cliver, 1983). The spectral signature of these flares is a continuum emission so short-lived that it is difficult to record (Grossi-Gallegos *et al.*, 1971; Rust, 1973). Easier is the observation of parts of the flare, generally small patches, seen in integrated light. The existence of the continuum emission indicates a dense flare plasma and thorough studies of this phase of larger flares provide valuable information on their nature.

In the history of morphological flare observations, two concepts which have played an important role in directing certain trains of research are *homologous* and *sympathetic* flares. With more than one active region visible on the Sun, flares may occur more or less simultaneously in different locations (Richardson, 1951) and there seems to be more of these so-called sympathetic flares than can be accounted for under the assumption of random occurrence of flares. Skylab data show that active regions are often connected by arch or loop structures, visible in soft X-rays and outlining magnetic field configurations. These coronal magnetic loops provide channels for different types of disturbances that can thereby travel from one active-region flare site to another. The concept of sympathetic flares thus leads us to consider the role played by magnetic fields. A similar emphasis on the importance of magnetic fields in the flare phenomenon is provided by the notion of homologous flares. It is often observed that a flare is brightening in exactly the same position and exhibiting the same geometrical outline as a previous flare in that region. This repetitive character was first discussed by Waldmeier (1938*b*). Such flares have been studied extensively, based on the point of view that the form of the magnetic field configuration is a determining factor in the processes that lead to a flare, and that there is a rebuilding of the stressed magnetic field after each successive flare. We are therefore again led to consider the role played by magnetic fields in flares.

It has been known for a long time that nearly all flares occur in active regions with sunspots, and the more magnetically complex the sunspot group, the higher the frequency of flare occurrence [e.g., Bell and Glazer (1959); Dodson-Prince and Hedeman (1970)]. This observation clearly points to the magnetic field as an important, maybe crucial, ingredient in the flare process. However, early observations showed that 'bigger is not better' in the sense that the flares are normally not found above the spot umbrae where the magnetic field is strongest (Švestka *et al.*, 1961). Evidently, aspects other than strength of the magnetic field configuration are important. On the other hand, when a flare occurs over a spot, new signatures appear as the flare plasma becomes the generator of radio microwave radiation and X-rays (Dodson-Prince and Hedeman, 1960; Ellison *et al.*, 1961; Martres and Pick, 1962). These early observations point to the hot, high-energy part of flares, the study of which has become increasingly important for our overall understanding of flares.

With the advent of solar magnetographs, direct comparison of flare position with pre-existing magnetic structures became possible (Bumba, 1958; Severny, 1958). Of particular interest is the location relative to the so-called 'magnetic neutral line' – the locus of points with zero longitudinal component of the magnetic field – a situation we shall further explore later. Martres *et al.* (1966) investigated this behavior in relation to the occurrence of small bright points of emission where the flare first occurs. As we shall see later, the study of these bright points has led to additional information on the flare phenomenon.

While most early magnetic field observations were restricted to measuring the longitudinal component, sporadic information was also gathered relating

relating to the transverse component [e.g., Zvereva and Severny (1970)]. During the big flare of August 1972, Zirin and Tanaka (1973) and Tanaka and Nakagawa (1973) discussed the importance of the observed sheared structure of the magnetic field in the region where the flare occurred. We shall see later that certain plasma motions that lead to sheared magnetic field structure do indeed seem to play a major role in flare production.

In the 1960s, our knowledge of solar flares increased dramatically with the Orbiting Solar Observatory (OSO) series of satellites which allowed us for the first time to study in detail the characteristics of flares at wavelengths inaccessible to ground-based observatories. These early spacecraft observations were soon supplemented by a host of others, such as from the European Space Research Organization TD-1A in 1972, NASA's Skylab Apollo Telescope Mount (ATM) in 1973–74 and, more recently, from the International Sun–Earth Explorer 3 (ISEE 3) and the US Air Force P78-1 satellite. Of particular interest are the data from NASA's Solar Maximum Mission (SMM) and the Japanese Hinotori spacecraft, with their emphasis on the high-energy aspects of solar flares. It is inappropriate here to discuss the vast amount of observational facts gathered from these spacecraft observations or from the host of rocket and balloon flights launched during the same period; we will have ample opportunity to discuss these observations throughout the course of the book.

We close this brief discussion by showing in Figure 1.1 the early time history of a fairly typical solar flare in various wavebands. We note that the manifestations of a flare are indeed complex and that, although there is some agreement between the light curves at different wavelengths, these manifestations are sufficiently diverse that no single one can fully describe the evolution of the flare. Added to this the fact that no spatial, polarization, or directionality information at any wavelength is presented in the diagram and that only a crude representation of a solar flare spectrum can be ascertained from it, one gains an idea of the degree of complexity of a solar flare and the potential wealth of information that we can extract from it.

1.2   Classification of flares

As with most complex phenomena, scientists have tried to devise schemes to classify flares and make some order of their seemingly chaotic and bewildering manifestations. The hope was that by systematically classifying certain observables, one would be able to gain some insight into the physics of the phenomenon.

1.2.1 *X-ray classification*

While all flare classifications prior to the 1960s have relied on observations in the visible part of the spectrum, more recent data suggest that the X-ray signature of flares may give as good, or better, criteria for classification purposes, so that we can deal with better coverage and rely on quantitative, more objective data. In addition, the X-ray criteria may simultaneously provide a more profound physical insight into the flare phenomenon itself.

Probably the simplest classification used is based on the global output of soft X-ray photons during a flare. This soft X-ray classification utilizes the flux in the 1–8 Å range of the spectrum, and, depending on the measured X-ray flux, one classifies a flare as a C, M, or X flare according to the scheme in Table 1.1. The classification letters represent the order of magnitude of the X-ray flux, and a subsequent numerical value indicates the multiple of the order of magnitude (e.g., $M3 = 3 \times 10^{-2}$ erg cm^{-2} s^{-1}). To accommodate flares smaller than class C, referred to as subflares, a class B has been added.

Fig. 1.1. Time profiles of the 22 GHz, hard X-ray, O V (1371 Å) and Fe XXI (1354 Å) emissions through the impulsive phase of a flare on 1 November 1980 (from Tandberg-Hanssen *et al.*, 1984).

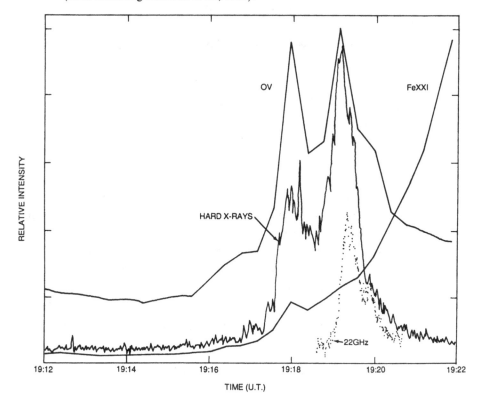

Table 1.1. *Soft X-ray flare classification*

Class	Intensity (erg cm^{-2} s^{-1})
B	10^{-4}
C	10^{-3}
M	10^{-2}
X	10^{-1}

Measurable X-ray levels below 10^{-4} erg cm^{-2} s^{-1} are indicated as decimal values of B (e.g., B.7 $= 7 \times 10^{-5}$ erg cm^{-2} s^{-1}).

While class C flares are a common occurrence during years near sunspot maximum, the frequency of X flares is always low. As an example, statistics (Smith, Jr, 1986) for the maximum years 1979–80 show that 2834 class C flares were observed, while the numbers for classes M and X were 554 and 59 respectively. For the minimum year 1976, only 75 class C flares were reported versus five class M flares. There was no class X flare. For the whole period 1976–85, there were reports of 17 986 flares, of which 74% were class C and 1% class X.

For most flares, a good correlation exists between the X-ray classification of Table 1.1 and the time-honored visual classification of Table 1.2, which we will discuss below.[1] Consequently, for these flares, not much is gained by changing (or adding to) the familiar visual classification scheme to the B, C, M, X notation of Table 1.1. However, some flares rate as very important in the X-ray domain, while their visual output would classify them as minor events – and vice versa for other flares. In these cases, we have added information that can tell us something about the nature of the flare in question, e.g., regarding the temperature of the radiating plasma, the location of the flare site, etc.

Furthermore, the X-ray signature can be used to describe a flare beyond giving its X-ray flux as indicated in Table 1.1. Several attempts have been made to use characteristics of the X-ray emission to classify flares in ways that reveal certain aspects of the flare physics. Time profiles of the flare emission at several frequencies (energies) enable us to divide flares into impulsive or gradual events and into flares with hard and/or soft X-ray spectra. From Skylab data, Pallavicini *et al.* (1977) studied soft X-ray flares seen at the limb and classified them into (*a*) compact flare loops, (*b*) point-

---

[1] This is not too surprising, and is an example of what has come to be known as the 'Big Flare Syndrome', which simply states that the bigger the flare, the brighter it is at all wavelengths.

like flares, and (c) large, diffuse flare loop systems. We shall return to some of these characteristics in Chapter 5.

Hard X-ray flares have been classified by Tanaka (1983) [see also Ohki *et al.* (1983) and Tsuneta (1983)] into three categories, viz.

Type A: thermal flares, in which a very hot ($T \approx 30$–50×10^6 K) plasma produces a smoothly varying thermal X-ray emission; these flares are small (< 5000 km) and compact and occur at low altitudes;

Type B: impulsive flares whose plasma, which is confined in long ($> 10^4$ km) sheared loops, produces impulsive spikes or bursts in the X-ray time profile;

Type C: gradual flares, whose plasma produces smoother, long-duration X-ray time profiles; these flares occur at high altitudes, $\sim 5 \times 10^4$ km, and the X-ray spectrum hardens with time.

Types B and C are generally considered nonthermal X-ray sources. We will return to this important point in Chapters 7 and 8.

1.2.2 *Visible light classifications*

A flare on the solar disk is observed as a temporary emission within some dark Fraunhofer line. The spectral line most frequently used is Hα, and in great flares the emission excess inside Hα may be several times the intensity of the adjacent continuum. On pictures of the disk, e.g., on a spectroheliogram or a Lyot filtergram, a flare is observed as a brightening of parts of the solar surface. The area covered by a great flare may be several times 10^9 km². Flares smaller than about 3×10^8 km² have generally been referred to as *subflares*.

It is this area that serves as the primary basis for visual light classifications. It is measured as the projected area in terms of the unit 1 square degree heliographic at the center of the disk, and should refer to the time of maximum brightness of the flare:

$$1 \text{ degree heliographic} = \tfrac{1}{360} \text{ solar circumference}$$

$$= 12\,500 \text{ km}.$$

One normally corrects the area for foreshortening and expresses it in millionths of the disk:

$$1 \text{ millionth of disk} = \tfrac{1}{97} \ (= 0.0103) \text{ square degree},$$

$$1 \text{ square degree} = 1.476 \times 10^8 \text{ km}^2 \text{ of solar surface}.$$

The above classification has the obvious drawback that it ignores the brightness of the flare. A unit area of flare emission is given the same weight whether it emits copious amounts of photons (i.e., is bright) or is a poor

Table 1.2. *Hα flare classification*

Corrected area		Relative intensity evaluation		
In square degrees	In millionths of hemisphere	Faint (f)	Normal (n)	Brilliant (b)
< 2.06	< 100	S f	S n	S b
2.06–5.15	100–250	1 f	1 n	1 b
5.15–12.4	250–600	2 f	2 n	2 b
12.4–24.7	600–1200	3 f	3 n	3 b
> 24.7	> 1200	4 f	4 n	4 b

emitter (i.e., is faint). To remedy this, a dual form for importance classification has been adopted (IAU, 1966). This importance evaluation consists of two elements, a number and a letter. The number describes the size of the area. The letter indicates whether the intensity of the flare area is faint (f), normal (n), or brilliant (b), and considerable subjectivity is associated with this part of the classification. The dual form for the importance classification is given in Table 1.2.

Radio observations have greatly added to our information of the flare phenomenon, and routine monitoring records flare emission at a number of frequencies. In general, the stronger the optical or X-ray emission from a flare, the stronger the radio flux, S. The latter is given in solar flux units, sfu, where $1\,\text{sfu} = 10^{-22}\,\text{Wm}^{-2} = 10^4$ Jansky. Using such flux measurements, we can assign a certain flux to the importance of a flare. As a rule of thumb, we may say that at 5 GHz, which corresponds to a radio wavelength of 6 cm, the microwave flux, S, from an optical Hα subflare is approximately 5 sfu, increasing to 30 sfu for a class 1, 300 sfu for a class 2, 3000 sfu for a class 3, and 30 000 sfu for a class 4. An approximate relation between importance class (Hα) and radio flux at 6 cm, expressed in sfu, is therefore

$$\text{Imp} = \log[S(\text{sfu})] - 0.5.$$

1.3   What does a flare look like?

It will be helpful for later discussions on flares to have in mind certain facts about the flare emission as observed in various wavelength regions. For a succinct overview see Bhatnagar (1986). Flares are roughly – and for practical purposes – divided into two groups (Pallavicini *et al.*, 1977), viz. small, compact flares and large, two-ribbon flares. However, we shall see later that it is not the distinction between compact, in the sense of small, versus two ribbons that is important from a physical point of view, but rather whether the flare is confined or not (Švestka, 1986). A compact flare probably takes place in a small loop in the lower corona, and the flare emission is largely confined to the plasma in the loop where it eventually dies away. The triggering, on the other hand, seems to be similar to the triggering of large flares, namely due to a large-scale eruption and reconnection of sheared magnetic fields (Sturrock *et al.*, 1984; Machado *et al.*, 1988a). By contrast, a two-ribbon flare is always associated with an erupting prominence, and the flare emission occurs in an arcade of 'post-flare loops' along the prominence with the individual loops oriented more or less at

Fig. 1.2. Development of a two-ribbon flare on 26 April 1979 following ejection of prominence material. Mhe panels at 20:03:14 and 20:03:20 UT show pictures taken in the blue wing of Hα where the ejected material is more visible (from Tang, 1986).

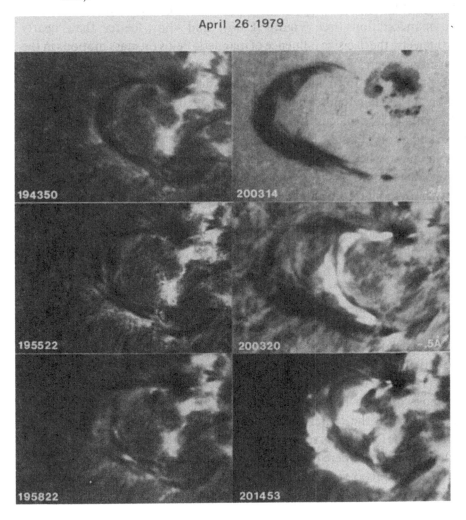

right-angles to the long axis of the prominence. Emission also occurs at the feet of the loops, thereby forming two ribbons on either side of the prominence. These ribbons of emission move apart as the flare progresses with larger and higher loops continuing to bridge them. The erupting prominence may eventually be confined or break away. In the latter case, we witness a coronal mass ejection (see Chapter 9). Figure 1.2 shows the development of a two-ribbon flare recorded in Hα. In this case, the field configuration allowed prominence material to escape to the outer corona, and a well-observed mass ejection resulted (Moore and LaBonte, 1980). Švestka (1986) refers to this class of flares as *dynamic*; see also Machado *et al.*, (1988a). During the development of two-ribbon flares the post-flare loops, when seen as loop prominences above the solar limb, exhibit material falling down from the loop apex along their legs and into the chromosphere. These post-flare loops are among the most distinguished features of large loop manifestations (see Figure 1.3(*a*)) and show better than nearly anything else the decisive influence of magnetic fields on mass motions in the solar atmosphere. Seen on the disk the loops – which then show up in absorption – again provide an interesting display as they connect the two ribbons of the flare emission (see Figure 1.3(*b*)).

Sometimes it is difficult to identify the two (or more) ribbons of a large flare, and at other times the ribbons exhibit exotic forms. One should keep this in mind when trying to model solar flares – often we make such simplified assumptions as to their morphology that we may invalidate the approach. An interesting and curious example of such a two-ribbon flare is shown in Figure 1.4 (Tang, 1985), where a loop-shaped, nearly circular prominence occupied the inversion line of magnetic polarity. One flare 'ribbon' filled part of the interior dominated by one polarity, while the other ribbon formed a bright band outside the filament in a region of the opposite polarity.

The smallest flares observed in Hα appear as unresolved bright blobs. Since larger flares that can be resolved generally show two footpoints (or two ribbons), it is reasonable to speculate that even the very smallest flares also have two footpoints, if they are flares in very small magnetic loops. Tang (1985) has been able to observe the footpoints in such small flares in the wings of Hα, where two distinct flare kernels can be detected, while observations in the center of Hα only reveal an unresolved bright blob, obscuring the footpoints; see Figure 1.5. The reason for this difference is that in the wings of Hα where the absorption coefficient is small (Section 2.1), we see down to the cooler chromospheric parts of the flare loop; whereas, in the center of Hα, we observe its higher-lying, hotter, and more disturbed parts.

Fig. 1.3. (*a*) Coronal loops of the post-flare type (courtesy Sacramento Peak Observatory, Air Force Cambridge Research Laboratories). (*b*) Post-flare loops seen in Hα in absorption, bridging the two bright Hα strands of a two-ribbon flare on 4 September 1982 (from Morishita, 1985).

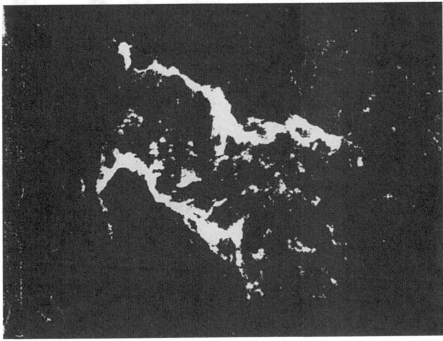

Fig. 1.4. Unusual two-ribbon flare of 11 September 1979 resulting in one nearly circular ribbon surrounding the second ribbon (courtesy F. Tang and Big Bear Solar Observatory).

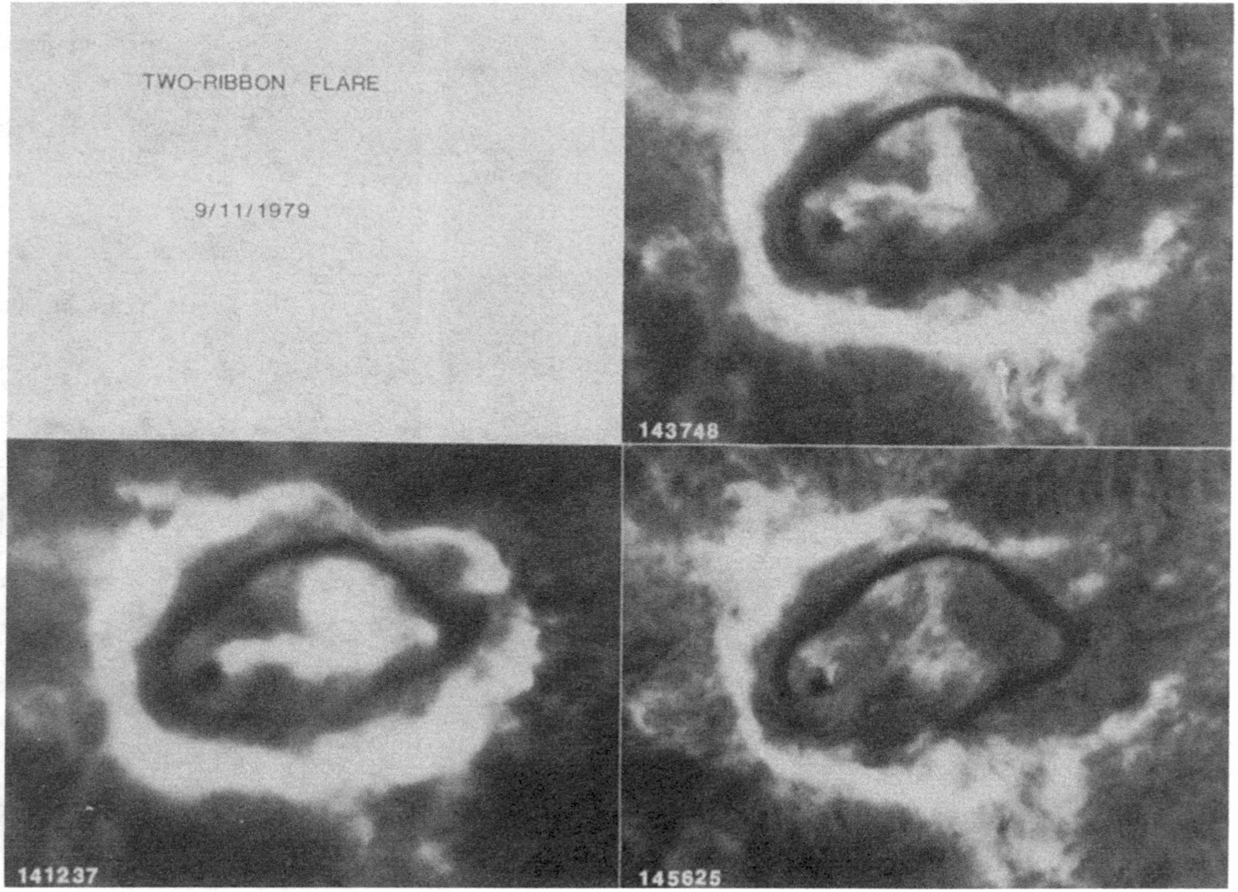

It follows from the above discussion that in Hα the emission may be considered as consisting of three parts, viz. bright kernels, extended areas (the ribbons), and post-flare loops connecting the ribbons. The total Hα emission from the kernels is about the same as from the ribbons. Zirin and Tanaka (1973) estimated the emission to be about 10^{27} erg s^{-1} in a big flare. The total Hα emission from post-flare loops is considerably less.

With the advent of high spatial resolution radio observations it became possible to image the flare plasma responsible for the emission of millimeter radio waves, thereby delineating the loop structure of another component of the flare plasma. Figure 1.6 shows the image of the microwave flare plasma and its relationship to the Hα flare.

Of greatest interest is the location of the high-energy flare plasma that produces hard X-ray emission and its spatial relationship to the cooler, low-energy Hα flare. Observations during the last solar maximum for the first time brought answers to questions concerning this problem. Figure 1.7 shows that a significant hard X-ray component originates in the feet of the flare loops, nearly co-spatial with the Hα emission; see also Kundu *et al.* (1984). In Chapter 7, we shall see how this mainly nonthermal emission can occur in such close proximity to the thermal Hα radiation. However, when we talk about proximity, we need to keep in mind that we cannot resolve the fine structure of the flare plasma, and only modeling can lead us to estimates

Fig. 1.5. Flare kernels visible in the red wing of Hα (Hα + 1 Å) but lost in the bright flare emission seen in the center of Hα from a compact flare on 15 June 1980 (courtesy F. Tang and Big Bear Solar Observatory).

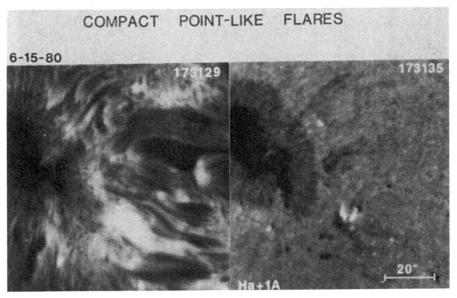

COMPACT POINT-LIKE FLARES

6-15-80

Fig. 1.6. Microwave flare plasma, observed at 15 GHz (white contours) and projected on an Hα map of the flare, is seen located in the top of a loop (or loops) in projection between the two loop footpoints: (*a*) at 15:15:52 UT and (*b*) at 15:18:52 UT, 8 September 1979. Notice the spread, with time, of microwave emission down the legs of the loop from its apex (from Marsh and Hurford, 1980).

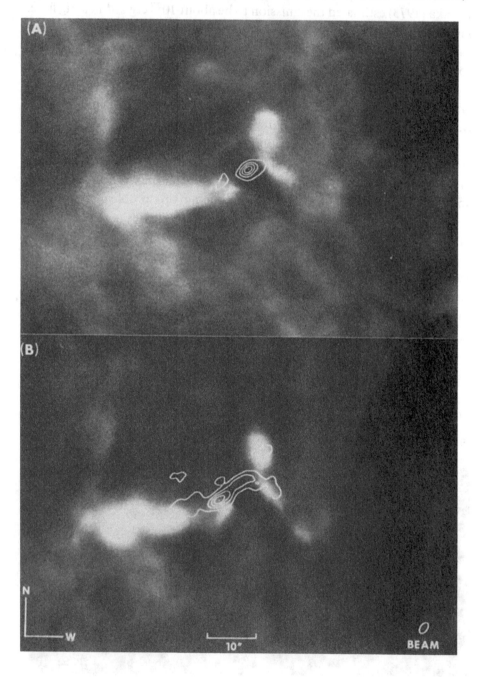

concerning the size of the hot and cool components of the plasma. We are led to this conclusion since there is considerable evidence that the physical processes in flares take place on scales perhaps orders of magnitude less than 100 km; in other words, in volumes much too small to be resolved with present-day observations. As an example of this evidence, Skylab observations of optically thin radiation from flares, which yield a quantity called *emission measure* (Section 2.6), were used together with an estimate of the electron density to derive a value for the characteristic length scale of the emitting plasma. By this method (see Section 8.1), estimates of electron densities in excess of 10^{13} cm^{-3} and characteristic scales less than 60 km

Fig. 1.7. Flare of 21 May 1980: (*a*) Hα image, location of hard X-ray emission (white contours) and prominence-filament (black); (*b*) line-of-sight component of the photospheric magnetic field; white is positive polarity, black negative; (*c*) Hα two-ribbon flare, with one ribbon in positive and the other in negative polarity; (*d*) line-of-sight component of photospheric magnetic field showing location of new emerging positive flux (*A*) and negative flux (*B*) prior to main flare development. The emerging flux was followed by filament eruption and flare onset (from Hoyng *et al.*, 1981).

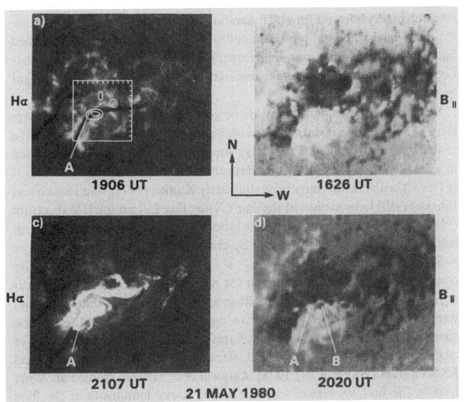

were obtained. Note that this region may have contained many energy-release regions on an even finer scale (Section 7.1).

It follows from this discussion that great care must be exercised when drawing conclusions from observations as to the real size of the flare plasma. Only a small fraction of the observed flare volume – the filling factor – may be filled with emitting flare plasma. Not before angular resolution of better than 0.1 arc sec (corresponding to spatial resolution on the Sun of better than 70 km) is reached in flare observations can we hope to resolve the structure of flare loops, and even then it is doubtful that we can resolve the size on which the fundamental physical processes take place. The answer to the question of how much better than 0.1 arc sec the resolution has to be is presently a matter of conjecture.

1.4   Stellar flares

Since the occurrence of flares is intimately connected with the behavior of the Sun's magnetic field in active regions (Chapter 6), one would also expect other stars with active-region fields to show flare activity. Due to its proximity, the Sun allows the study of its flares to be much more detailed than any study of stellar flares, and this circumstance makes solar flare research highly relevant for understanding flaring stars. On the other hand, the often vastly different physical conditions on flaring stars can help shed light on processes only marginally operative in solar flares. Little use has been made so far by solar flare researchers of this potentially important 'solar–stellar connection'.

More than 30 years ago some dMe objects were identified as the first 'flare stars'. These red dwarf stars show sudden and short-lived brightenings in the visible spectrum, indicating that large amounts of energy have been released in a flare-like process. The optical flare characteristics of Proxima Centauri (α Cen C), a dM5e star, have been studied by Kunkel (1973), and Haisch and Linsky (1980) have measured the star's X-ray flux L_x and the UV spectrum (1175–3200 Å) outside of flares. The latter authors found that the M dwarf has a quiescent corona that can be characterized by a temperature of $\sim 3.5 \times 10^6$ K, $L_x = 1.5 \times 10^{27}$ erg s^{-1}, and $L_x/L_{bol} = 2.2 \times 10^{-4}$. By comparison, the quiet-Sun value for the ratio of X-ray to bolometric flux is $L_x/L_{bol} = 1.3 \times 10^{-6}$. Vaiana and Rosner (1978) estimated that if the Sun were completely covered with active regions, $L_x/L_{bol} \approx 5 \times 10^{-5}$. While some of the stars that produce X-ray flares are dMe dwarfs, others belong to 'active' stars such as young giant stars, pre-main sequence stars, and RS CVn stars (Walter *et al.*, 1978). Agrawal *et al.* (1986a) studied an X-ray flare in the RS CVn binary star σCrB. The X-ray luminosity at the flare maximum was 6×10^{30} erg s^{-1} and the total energy radiated in the X-ray band was 2×10^{34} erg. For these active RS CVn stars, $L_x/L_{bol} = 10^{-3}$ in their quiescent phase (Agrawal *et al.*, 1986*b*). Montmerle *et al.* (1983) suggested that very active objects like T Tau stars may show a continuous X-ray flaring activity, and Skumanich (1985) has suggested that the flare mechanism in dwarf stars operates both on a large scale (producing flares) and on a small scale (producing microflares), and that microflares give rise to a quiescent X-ray corona.

Some flare stars also show activity in the radio frequency domain. Lang *et al.* (1983) studied the dMe star AD Leo at 20 cm wavelength and found radio flares, short energetic spikes of duration less than 1 s [see also Lang (1986)], while Jackson *et al.* (1980) found longer lasting, large flux increases at 6 and 20 cm in the dMe star UV Ceti.

In all these cases suggestions have been made that stellar magnetic fields are at the root of the flare activity, and we shall see in Chapter 6 that magnetic fields are also responsible for solar flares. Zirin and Ferland (1980) drew attention to observations that clearly show how solar and stellar flares are similar in nearly all respects, except that stellar flares can be much more spectacular emitters, especially in the bluer parts of the spectrum. The imposing energy displays in some flaring dMe stars may lead to emissions 10^6 times solar flare values.

Special interest is attached to the so-called blue continuum observed in some stellar flares (Mochnacki and Zirin, 1980). Stark broadening (see Section 8.2.2.1) of hydrogen lines seems to contribute to the blue continuum emission in solar flares. The same mechanism has also been proposed for stellar flares by Zarro and Zirin (1985). They compared the appearance of the hydrogen lines during a flare on the star YZ CMi with their appearance during quiescent conditions and showed that at least part of the flare continuum shortward of 3800 Å may be due to merging of the broadened hydrogen lines.

1.5   Terrestrial effects

It is outside the scope of this book to discuss in any detail the effect of flares on the terrestrial environment. However, for the sake of completeness and in order to fathom the physical complexity of larger flares, we shall briefly look at those flare agents that most significantly influence the Earth and its immediate environment.

1.5.1   *High-energy photons*

At times, flares release copious amounts of X-rays which, in turn, cause extra ionization in the Earth's ionosphere. This increased electron density is responsible for absorption and refraction of radio waves propagating from one station, through the ionosphere, to another station. Flare forecasting can therefore have practical importance in the realms of short-wave radio communication in that sudden loss of signal or abrupt change in transmission paths may be anticipated.

1.5.2   *Energetic protons*

In sufficiently violent flares, acceleration of energetic protons takes place. When these protons reach the Earth's atmosphere they enhance the ionization, especially the polar ionosphere, where their entry is facilitated along the geomagnetic field lines. Of practical importance is the resulting increase in absorption of high-frequency waves propagating through this region. Also, the energetic protons, by direct collision, may damage electronic components in satellites and may be a health hazard to space travelers.

1.5.3   *Shock waves*

The existence of a flare-induced disturbance of the Earth's magnetic field was suspected as far back as 1859, when a severe geomagnetic storm followed the now legendary white-light flare observed by Carrington (1859). Note, however, that even Carrington himself was somewhat reluctant to admit the association, noting, after Aristotle, that 'one swallow does not a summer make'. Such sudden commencement (s.c.) storms and their associations with flares are now well understood. The storms are caused by shock waves, which are released in the flare, and travel through space at approximately 100 km s^{-1}, even though speeds up to 3000 km s^{-1} have been observed [see Hundhausen (1972a, b); Dryer (1974)]. While many s.c. storms are caused by solar wind discontinuities unrelated to flares (Gosling *et al.*, 1967), the most clearly defined s.c. storms owe their existence to shocks originating in flares, and Burlaga and Ogilvie (1969) have shown that these are hydromagnetic shocks.

As the shock waves move out from the Sun, not only will they affect the terrestrial environment, but will collide with other planets as well. Dryer *et al.* (1982) reported on an interplanetary shock wave that was produced during a solar flare and observed to interact with the atmosphere of Venus where it produced a significant compression of the dayside ionosphere.

An important terrestrial effect, the *Forbush decrease*, takes place when the shock waves sweep across the geomagnetic field lines and deflect the galactic cosmic ray particles that constantly bombard the Earth, thereby causing a decrease in the particle count (Haurwitz *et al.*, 1965; Švestka, 1966b).

1.5.4 *Coronal mass ejections*

As we shall see in Chapter 9, flares are often associated with expulsion of matter from the overlying corona. These coronal mass ejections may manifest themselves as magnetic clouds or bubbles when they reach the Earth, where they cause changes in the equatorial D_{st} geomagnetic index and

Fig. 1.8. Superposed epoch analyses of D_{st} and B_z for 19 magnetic clouds showing their geomagnetic effect (after Wilson, 1987).

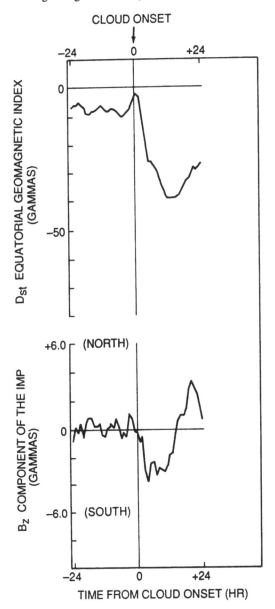

the B_z component of the interplanetary magnetic field. The effect was first suggested by Burlaga (1971) and studied later by Klein and Burlaga (1982); Crooker (1983); and Wilson and Hildner (1984, 1986). Figure 1.8, after Wilson (1987), shows the superposed epoch analyses of B_z and D_{st} for 19 magnetic clouds observed between 1973 and 1978. The geomagnetic effect of the magnetic clouds is clearly discernible.

This completes our brief overview of the observational characteristics of solar flares. In order to address the physical processes occurring in, and responsible for, them, it is necessary to use some fundamental results from certain areas of physics. The next four chapters are, therefore, concerned with the development of these required tools.

2

Flare spectroscopy

A few strong instincts, and a few plain rules
Alas! What Boots the Long Laborious Quest? W. Wordsworth (1770–1850)

A wealth of observations is available on the electromagnetic spectrum of different types of solar flares, ranging from photons with energies less than 10^{-6} eV (i.e., meter radio waves) to photons with energies exceeding 10 MeV (i.e., hard X-rays and γ-rays). These data carry information that can provide diagnostic tools to determine physical parameters of the flare plasma that are crucial for an understanding of the physics of solar flares. In this chapter we shall review basic spectroscopic concepts and discuss different radiation mechanisms likely to be found operative in flare plasmas.

For many years there has existed a vast amount of literature on flare spectroscopy as it relates to the visible part of the spectrum. Although we shall refer heavily to it, we emphasize that this is only an example of application of a technique to one part of the electromagnetic spectrum. Equally important information can be similarly derived from the other parts, i.e., X-rays, γ-rays, UV, infrared, and radio waves.

Depending on the methods we use to observe flare radiation we encounter photons generated in the flare plasma by a variety of mechanisms, e.g., Bremsstrahlung, electron transitions in atomic structures, synchrotron processes, Compton scattering, nuclear reactions, etc. Prior to 1970, by far the bulk of flare spectroscopy involved the optical part of the spectrum where free–free and free–bound electron transitions in atoms and ions under nonrelativistic conditions dominated. With the increased emphasis on hard X-ray, γ-ray, and radio microwave observations in the late 1970s – and especially during the Solar Maximum Mission – an exciting body of data is now available concerning photons emitted in nuclear reactions, particle annihilations, electron acceleration in magnetized plasmas, etc.

2.1 Concepts in radiation transport

As radiation of frequency ν travels a distance ds, its specific intensity, I_ν (erg cm^{-2} s^{-1} Hz^{-1} sr^{-1}), will change according to the expression

$$dI_\nu = \varepsilon_\nu \, ds - \kappa_\nu I_\nu \, ds, \tag{2.1}$$

where ε_ν and κ_ν are the emission and absorption coefficients per unit volume. We can define an optical depth τ_ν along a direction r that makes an angle θ with ds

$$d\tau_\nu = -\kappa_\nu \, dr, \tag{2.2}$$

and find the transport equation

$$\mu \frac{dI_\nu}{d\tau_\nu} = I_\nu - \frac{\varepsilon_\nu}{\kappa_\nu}, \tag{2.3}$$

with $\mu = \cos \theta$.

The ratio $\varepsilon_\nu/\kappa_\nu$ defines the *source function* S_ν. When local thermodynamic equilibrium reigns, the ratio $\varepsilon_\nu/\kappa_\nu$ depends only on the temperature T (Kirchhoff's law), and the source function is the Planck function $B_\nu(T)$; i.e.,

$$S_\nu = B_\nu(T) \equiv \frac{2h\nu^3}{c^2} \left[\exp\left(\frac{h\nu}{kT}\right) - 1 \right]^{-1}, \tag{2.4}$$

where h is Planck's constant, k is Boltzmann's constant, and c is the velocity of light.

For radio waves $h\nu \ll kT$, and Planck's law simplifies to the classical Rayleigh–Jeans law

$$B_\nu(T) = \frac{2kT}{c^2} \nu^2. \tag{2.4'}$$

Instead of using frequency one often expresses the intensity in terms of wavelength: I_λ (erg cm^{-2} s^{-1} cm^{-1} sr^{-1}), $\lambda = c/\nu$, and the corresponding form of Planck's function is

$$B_\lambda(T) = \frac{2hc^2}{\lambda^5} \left[\exp\left(\frac{hc}{\lambda kT}\right) - 1 \right]^{-1}. \tag{2.4''}$$

Equations (2.1) through (2.4) form the basis for discussion of transport of continuum radiation. In the case of spectral lines, especially strong lines, the source function may depart significantly from the Planck function, resembling more the direction-averaged mean intensity J_ν [see Chandrasekhar (1950)]:

$$J_\nu = \frac{1}{4\pi} \int I_\nu \, d\Omega. \tag{2.5}$$

For such spectral lines we need a more general approach to the problem of a radiating plasma than that provided by the conditions of local thermodynamic equilibrium (LTE), and we replace these conditions by the assumption of a statistically steady state in each atomic energy level. When we include collisionally induced transitions this treatment is a generalization of the principle of reversibility, already discussed by Rosseland (1926) and developed by Thomas (1948a, b) and others.

Let us consider an atom where a line is formed by transition between two bound levels i and j, where $j > i$, and let κ denote the continuum. A general form of the line source function is (Thomas, 1957)

$$S_{ji} = \frac{2h\nu_{ij}{}^3}{c^2} \left(\frac{n_i g_j}{n_j g_i} - 1 \right)^{-1} \frac{\psi_\nu}{\Phi_\nu},$$ (2.6)

where g_i and g_j are the statistical weights of levels i and j, respectively, and Φ_ν and ψ_ν are the profiles of the absorption and spontaneous emission coefficients. We shall discuss equation (2.6) more later, but we notice here that the ratio of the level populations is required.

Let P_{ij} denote the total transition rate from level i to level j. This rate is composed of a radiative part, R_{ij}, and a collisional part, C_{ij},

$$P_{ij} = R_{ij} + C_{ij}.$$ (2.7)

Similarly, $P_{ji} = R_{ji} + C_{ji}$ and R_{ji} will then be the Einstein coefficient for spontaneous emission, A_{ji}, plus the transition rate due to induced (or stimulated) emission, $B_{ji} \int J_\nu \bar{\psi}_\nu \, d\nu$, where B_{ji} is the Einstein coefficient for induced emission, J_ν the mean intensity [equation (2.5)], and $\bar{\psi}_\nu$ the profile of the induced emission coefficient. Let us now define the emission–coefficient-weighted mean intensity $\bar{J}$ through

$$B_{ji} \int J_\nu \bar{\psi}_\nu \, d\nu = \bar{J} B_{ji},$$

so that

$$R_{ji} = A_{ji} + \bar{J} B_{ji}.$$ (2.8)

The B_{ji} coefficients are computed from the A_{ji} coefficients according to the formula

$$B_{ji} = \frac{c^2}{2h\nu^3} A_{ji}.$$ (2.9)

Similarly, the radiative transition rate for absorption is

$$R_{ij} = B_{ij} \int J_\nu \phi_\nu \, d\nu = \bar{J} B_{ij},$$ (2.10)

where B_{ij} is the Einstein coefficient for absorption and is related to the coefficient for stimulated emission by

$$\frac{B_{ij}}{B_{ji}} = \frac{g_j}{g_i}.$$ (2.11)

Here ϕ_ν, the profile of the absorption coefficient, is normally assumed to be equal to ψ_ν.

Finally, we need to consider the transition rates to and from the continuum. The photo-ionization rates are

$$R_{i\kappa} = 2\pi \int \frac{\sigma_\nu J_\nu \, d\nu}{h\nu} \left(= \frac{2\pi}{hc} \int \sigma_\lambda J_\lambda \lambda \, d\lambda \right), \tag{2.12}$$

where σ_λ is the ionization cross-section, often calculated following a method based on the quantum defect (Burgess and Seaton, 1960). The rates from the continuum, i.e., the photo-recombination rates, $R_{\kappa i}$, are often calculated by balancing photo-ionization and recombination in LTE, which is assumed to hold in the continuum, i.e.,

$$n_\kappa^* R_{\kappa i} = n_i^* B_{i\kappa} B(T_e). \tag{2.13}$$

Here an asterisk denotes the value of the population in LTE and $B(T_e)$ is the Planckian radiative field at the electron temperature T_e [equation (2.4)]. In equilibrium the simple ionization formula due to Saha can be used

$$\frac{n_\kappa^*}{n_i^*} = \frac{(2\pi mkT)^{3/2}}{n_e h^3} \frac{U_k}{U_i} \exp\left(-\frac{\chi_k - \chi_i}{kT_e} \right). \tag{2.14}$$

The partition functions, U, measure the number of electrons in all levels according to Boltzmann's formula

$$U_i = \sum_{s=1}^\infty g_{i,s} \exp\left(-\frac{\chi_s}{kT_e} \right), \tag{2.15}$$

and $g_{i,s}$ is the statistical weight of the sth level of the ith state.

In addition to radiation, collisions also help populate, and depopulate, levels i and j as well as the continuum κ. The collisional rate is

$$C_{ij} = \int_{v_0}^\infty Q_{ij} n_e v f(v) \, dv, \tag{2.16}$$

where the integral is taken from a threshold velocity v_0 of the electrons whose velocity distribution function is $f(v)$. Here Q_{ij} is the cross-section for collisional excitation; similarly for ionization $Q_{i\kappa}$.

Collisional de-excitation rates are formed by balancing them against the excitation rates in LTE

$$\frac{C_{ji}}{C_{ij}} = \frac{n_i^*}{n_j^*} = \frac{g_i}{g_j} \exp\left(\frac{\chi_i - \chi_j}{kT_e} \right), \tag{2.17}$$

i.e., the ratio of the population is given by Boltzmann's law. We shall return to the special conditions of dielectronic recombination and auto-ionization later.

We can now go back to equation (2.7) and calculate how the various levels of an atom, or ion, are populated, and form the ratio n_i/n_j necessary to work with the source function (2.6). The ratio n_i/n_j must be taken from the solution of the statistical equilibrium equations for the levels, i.e.,

$$\frac{dn_j}{dt} = \sum_{i=1,i\neq j}^{\kappa} (n_i P_{ij} - n_j P_{ji}) = 0, \quad j = 1, 2, \ldots, \kappa. \tag{2.18}$$

With the definition

$$\sum_{i=1,i\neq j}^{\kappa} P_{ji} = P_{jj}, \tag{2.19}$$

equation (2.18) may, in a steady state, be written

$$\sum_{i=1,i\neq j}^{\kappa} (n_i P_{ij}) - n_j P_{jj} = 0, \quad j = 1, 2, \ldots, \kappa. \tag{2.20}$$

The algebraic solution of (2.20) can be written (White, 1961) [dating back to Rosseland (1936)]

$$\frac{n_i}{n_j} = \frac{P^{ij}}{P^{ii}},$$

where P^{ij} is the co-factor of the element P_{ij} in the matrix of the coefficients of equation (2.20). This form of the solution is particularly convenient for use in the source function (2.6).

However, we now come to the circumstance that makes the radiative transfer problem of strong lines so complicated. Both the source function S_{ji}, as well as several of the transition rates P_{ji}, contain the mean radiation field J_ν [equation (2.5)], which itself is given by the transfer equation containing the source function; see equation (2.3). A more appropriate form of the transfer equation when dealing with lines may be obtained as follows: first integrate equation (2.3) over all directions μ to obtain

$$\frac{1}{2\pi} \frac{dF_\nu}{d\tau_\nu} = 2(J_\nu - S_\nu),$$

where $F_\nu = 2\pi \int_{-1}^{1} I_\nu \mu \, d\mu$ is the *radiation flux* (erg cm^{-2} s^{-1} Hz^{-1}). Then multiply (2.3) by μ and again integrate over all directions. Using the approximation $\int_{-1}^{1} I_\nu \mu^2 \, d\mu = \bar{I}_\nu \int_{-1}^{1} \mu^2 \, d\mu \approx \frac{2}{3} J_\nu$ (thereby equating two somewhat different measures, J_ν and $\bar{I}_\nu$, of the average radiation field), we find

$$\frac{2}{3} \frac{dJ_\nu}{d\tau_\nu} = \frac{F_\nu}{2\pi}.$$

Elimination of F_ν between the last two equations gives the *Eddington approximation*

$$\frac{1}{3} \frac{d^2 J_\nu}{d\tau_\nu^2} = J_\nu - S_\nu, \tag{2.21}$$

where τ_ν is the total optical depth, line plus continuum, and the source function S_ν is a function of both the line source function S_{ji} and the continuum source function S_κ. As noted above, S_κ may be taken equal to the Planck function [equation (2.4)]. The total source function S_ν is defined as

$$S_\nu = \frac{\varepsilon_{ji} + \varepsilon_\kappa}{\kappa_{ji} + \kappa_\kappa},$$

or, by introducing the ratio $r_\nu = \kappa_\kappa / \kappa_{ji}$,

$$S_\nu = \frac{S_{ji} + r_\nu S_\kappa}{1 + r_\nu}.$$

When we insert this expression for the source function in equation (2.21), we arrive at an integro-differential equation for J_ν, and Jefferies and Thomas (1958) showed that an analytic solution is possible in terms of a Gaussian quadrature over frequency.

2.2   Interaction of photons with atoms and electrons

In equation (2.1) we introduced the specific intensity I_ν and discussed its transport through a plasma. We shall now look a little closer at this radiation and see how it is described in the language of quantum mechanics.

We consider an optically thin emission line, i.e., a case where τ_ν of equation (2.21) is small. The emergent intensity of such an emission line is given by

$$I_{ji} = n_j A_{ji} h\nu_{ji} \, dV, \tag{2.22}$$

where n_j is the population of the upper atomic level from where the electrons jump to the lower level i during the emission of the quanta (photons) $h\nu_{ji}$, and V is the volume of the plasma radiating the line ν_{ji}. The population, n_j, of an energy level j in an atom (or ion) depends on

(i) the ratio, n_j/n_{ion} of n_j, to the total number of atoms in that particular ionization state, n_{ion},

(ii) the ionization equilibrium, n_{ion}/n_{el}, where n_{el} is the elemental abundance,

(iii) the abundance of the element relative to hydrogen, n_{el}/n_H, and

(iv) the hydrogen density, n_H.

Hence we may write for n_j,

$$n_j = \frac{n_j}{n_{ion}} \frac{n_{ion}}{n_{el}} \frac{n_{el}}{n_H} n_H. \tag{2.23}$$

Only under thermodynamic equilibrium conditions can we find n_j/n_{ion} and n_{ion}/n_{el} by simply applying Boltzmann's and Saha's formulae. More generally, to interpret the information contained in flare line spectra correctly, we need a closer study of the population of the energy levels involved.

The energy levels in an atom or ion – which indicate where an electron can be bound – are characterized by their quantum numbers: n (principal), s (spin), l (angular momentum), and j (total angular momentum, i.e., coupling of spin with orbital angular momentum). The principal quantum number n reflects the ordering of the levels at increasing orbital radii (as in the simple Bohr picture) or with increasing energy, E_n. A photon is emitted (or absorbed) by the atom when the electron jumps from one energy level (n, s, l, j) to another (n', s', l', j'). If there is more than one electron in an atom, one uses the lower-case symbols to denote the angular momenta for a certain electron and capital letters L, S, and J to symbolize the momenta of the entire atom. In this case with more than one electron, they are arranged in levels (or 'shells'), always obeying Pauli's exclusion principle which states that no two electrons may have the same quantum numbers. The levels (n, S, L, J) are designated, for historical reasons, by a letter $(S, P, D, F, G,$ etc.) according to the value of L (i.e., $L = 0$ is called S, $L = 1$ is P, etc.), by a superscript giving the multiplicity, $2S + 1$, which indicates how many J values are available, and by a subscript giving the J value. The value of n is placed in front of the letter. For example, the ground level of the helium atom is designated 1^1S_0.

Let the energy level (n, S, L, J) denote an upper level, j, in our previous notation. Similarly, let the level (n', S', L', J') be a lower level, i. From a plasma consisting of atoms with these levels we will, if an electron jumps from j to i, observe a spectral line of frequency

$$v_{ji} = \frac{E_j - E_i}{h}, \tag{2.24}$$

where E_j and E_i are the energies of the levels j and i respectively. An emission line occurs for a jump from level j to level i, while a jump $i \rightarrow j$ results in an absorption line, since in that case the photon is absorbed by the atom and its energy is used to raise the electron to the higher level.

One of the fascinating things about spectroscopy is that it is possible to predict what electron jumps – transitions – will occur under given conditions, and thereby understand the ensuing set of lines, i.e., the resulting atomic spectrum. The reason for this predictability lies in the existence of certain rules obeyed by the quantum numbers. When an electron jumps, it

can do so only if the change in quantum numbers is consistent with the
following restraints:

$$S' = S \text{ (i.e., spin is preserved)}, \quad \Delta S = 0,$$
$$L' = L \text{ or } = L \pm 1, \qquad\qquad \Delta L = 0, \pm 1, \qquad (2.25)$$
$$J' = J \text{ or } = J \pm 1, \qquad\qquad \Delta J = 0, \pm 1.$$

There is no restriction on n. In addition, the following rules must be obeyed:

The transition from $J = 0$ to $J' = 0$ is forbidden, and parity must
change by 1, where parity is defined as the sum of the l-values of all
the electrons.

2.2.1   *Degeneracy of atomic levels*

When we talk about a spectroscopic level, we understand an
electron position in the atom determined by the four quantum numbers, n,
S, L, and J, e.g., $2^2P_{3/2}$. A transition between two levels results in a spectral
line. If one applies a magnetic field to the radiating atom, quantum theory
shows that the total angular momentum J may be pointed only in certain
'quantized' directions with respect to the magnetic field. In particular, J may
be pointed such that its projection along the magnetic field can take one of
the following values, and no other: $J, J - 1, \ldots, -J$ or in all $2J + 1$ values.
This projection, called M_J and furnishing a fifth quantum number, indicates
that a level is composed of $2J + 1$ sublevels that often are indistinguishable.
We say the level is *degenerate*, and the degree of degeneracy is the number
$2J + 1$ which is also called the *statistical weight* of the level; i.e., $g = 2J + 1$.
It is basically the number of electrons that can occupy a level without
violating Pauli's exclusion principle. The sublevels, or states, into which a
level is split, are now designated by the quantum numbers n, S, L, J, and M_J.
It turns out that the stronger the applied magnetic field, the larger the
separation of the states. In Figure 2.1, the spectral line is indicated by the
transition between the (degenerate) levels $J = 2$ and $J = 1$, and the
components that make up the line are shown by the transitions between the
states, each characterized by its magnetic quantum number M_J. We notice
that not all states communicate with each other, only those for which the
rule for M_J applies; i.e.,

$$M_J' = M_{J'} \text{ or } M_J \pm 1, \quad \Delta M_J = 0, \pm 1. \qquad (2.25')$$

In addition to considering a line and its possible components, the concept
of a *multiplet* is useful in spectroscopy. A multiplet is the collection of all
possible spectral lines between all levels with the same n, S, and L quantum
numbers, but with different values of J. A collection of such levels is called a

term. Knowledge of what lines belong to a given multiplet often has more than academic interest. For example, when a line L_1, useful in density diagnostics (Section 8.2.1.2), is blended with a line L_2 of another element, it may not be possible to assess the strength of line L_1, thereby making a density determination impossible. However, if we can observe other lines of the multiplet to which L_1 belongs, reasonable estimates of the strength of L_1 may be possible.

2.2.2 *Forbidden lines*

In the classical treatment the radiation field can be expanded in a series where the first and dominant term is the electric dipole contribution. If other terms are neglected, this gives a good first approximation which corresponds, in the quantum theory, to the electron jumps governed by the quantum number rules in equations (2.25) and (2.25'). However, similar to the fact that in classical theory we have secondary terms in the expansion for the radiation field (i.e., electric quadrupole and higher terms and magnetic dipole and higher terms), in the quantum picture of the atomic structure we encounter transitions that do not obey the rules of equation (2.25), but that obey other sets of rules and may lead to observable spectral lines. Such transitions may not lead to the most important flare lines, but they are of

Fig. 2.1. Schematic representation of a spectral line between two degenerate levels ($J = 2$ and $J = 1$) and its allowed components between states designated by the M_J quantum numbers, when the degeneracy of the levels is lifted due to the application of a magnetic field.

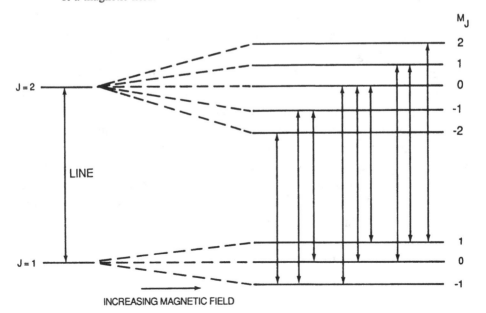

prime interest in coronal physics, where nearly all the strong lines are due to magnetic dipole radiation. On the other hand, electric quadrupole radiation is rarely important in solar physics.

An important example for solar physics is the so-called 'green line' of Fe XIV at 5303 Å. This line results from a transition from an upper level which is metastable and is, under laboratory (high-density) conditions, rapidly depopulated by collisional excitation to an even higher level, with subsequent decay from that level. However, under the low-density conditions appropriate to, for example, the solar corona, this upward collisional excitation occurs at a rate slower than the magnetic dipole de-excitation from the metastable level, so that the emission in the green line becomes significant.

The selection rules for magnetic dipole radiation are:

$$\left.\begin{aligned} S' &= S, \\ L' &= L, \\ J' &= J \text{ or } J \pm 1, \\ n'l' &= nl; \end{aligned}\right\} \tag{2.26}$$

i.e., the transitions leading to magnetic dipole radiation take place between levels within one and the same term, i.e., $^3P_{3/2} \rightarrow {}^3P_{1/2}$. The corresponding transition probabilities $A_{J' \rightarrow J}$ are about 10^5 times smaller than for electric dipoles.

2.2.3 *Intermediate coupling*

When the electron spin does not interact too strongly with the orbital angular momentum, the quantum numbers S and L adequately describe the energy states. This situation is referred to as LS coupling. However, in certain atomic structures the spin–orbit interaction perturbs the atomic states to a degree that the states partially mix with one another, and the quantum numbers S and L do not suffice to describe this coupling which is referred to as intermediate coupling. One of the simplest and best-known examples of this coupling is found in the neutral helium atom where a partial mixing takes place between the 2^1P_1 and 2^3P_1 terms. Normally a ^3P term cannot communicate with a ^1S term, but in helium $^3P \rightarrow {}^1S$ transitions become possible because $^3P_1$ has some $^1P_1$ mixed with it in this intermediate coupling case. As a consequence, a line at 591 Å due to the $^3P \rightarrow {}^1S$ transition is observed. In highly ionized helium-like ions the spin–orbit interaction may lead to considerable splitting between the J levels (J–J coupling), with the mixing becoming so strong that important spectral lines ensue.

2.2.4   *Inner shell transitions*

In the previous discussion of spectral line formation, we have tacitly assumed that it is the outer (valence) electron that is undergoing the jump in quantum number(s). However, under certain conditions an inner shell electron may be involved, resulting in the emission of a line in the so-called characteristic X-ray spectrum of the element.

Figure 2.2(a) shows schematically the energy levels of a chemical element. The K, L, M, N, etc., shells are filled with their prescribed (by Pauli's exclusion principle) numbers of electrons (2, 8, 18, etc.). At the top are the unfilled shells (the 'optical' levels; electron jumps to these levels result in the optical spectrum). A very energetic electron or photon ($> \sim 10^3$ eV) can remove one of the two K-electrons either to a higher unfilled level

Fig. 2.2. (*a*) Schematic Bohr picture of electron orbits, or shells, and transitions to the two innermost shells. (*b*) Schematic representation of the Kα_1 and Kα_2 transitions between the L- and K-shells.

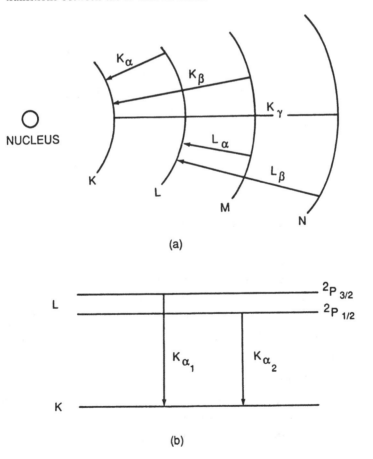

(a)

(b)

(excitation) or to the continuum (ionization). When an electron from the L-shell makes a transition to the K-shell to fill the vacancy, a Kα photon is emitted; its energy places the corresponding emission line in the X-ray part of the spectrum. Similarly, if an M-shell electron jumps to the K-shell, we obtain a Kβ line, etc. The complete group of lines forms the K-series.

One of the best studied cases in flares pertains to the Kα_1 and Kα_2 lines in Fe II at 1.936 Å and 1.940 Å respectively. The Kα_1 line corresponds to de-excitation from the $2P_{3/2}$ level and the Kα_2 line to de-excitation from the $2P_{1/2}$ level; see Figure 2.2(b). On statistical weight arguments, we therefore expect the Kα_1 line to be twice as intense as the Kα_2 line; this has been verified in actual observations of solar flares (Culhane *et al.*, 1981). As mentioned, a K-shell electron may be removed either by radiation (fluorescence excitation) or by collisions (electron impact ionization), and we shall see later, in Section 7.6.1, that studies of the resulting lines provide important diagnostics for the nature of the flare energy-release mechanism by setting an upper bound on the collisional ionization rate.

2.2.5   *Dielectronic recombination and auto-ionization*

Burgess (1964) showed that under certain conditions (e.g., a hot, tenuous plasma as found in the solar corona) dielectronic recombination is as important as ordinary radiative, two-body recombination. Even though Burgess apparently overestimated the coefficient for dielectronic recombination (Alam and Ansari, 1985), the process still competes favorably with ordinary radiative recombination.

The difference between the two processes is that while in ordinary recombination the electron is simply captured with the release of a free–bound photon, in dielectronic recombination we first get a doubly excited unstable ion as the captured electron, which ends up in an excited level, excites another electron to another excited level. In other words, we end up in this first step with a doubly excited ion of one less positive charge. For this to happen, the incoming electron must have just the right energy to excite the other electron, and the incoming electron itself is captured into an excited level above the first ionization limit (normally a high Rydberg state). This process, for an ion N with charge Z can be described by the formula

$$N^{+Z}(n_i l_i) + e^- \rightarrow N^{+(Z-1)**}(n_f l_f, n''l''), \qquad (2.27)$$

where n and l are the quantum numbers of the levels, with $n''l''$ indicating the captured electron. The excited levels above the first ionization limit are referred to as auto-ionization levels of the lower ion and mingle with the bound levels of the higher ion. Reaction (2.27) can therefore easily go in the reverse direction; i.e., the outer electron will auto-ionize. However, if the inner electron makes a downward transition, the excess energy is radiated away. The ion is stabilized, and we have an outgoing photon and an ion of charge $+(Z + 1)$, according to the following reaction

$$N^{+(Z-1)**}(n_f l_f, n''l'') \rightarrow N^{+Z*}(n_i l_i, nl) + hv. \tag{2.28}$$

Maximum transition probability occurs for $n''l'' = nl$ [see, e.g., Geltman (1985)].

Since the ion $N^{+(Z-1)**}(n_f l_f, n''l'')$ can either auto-ionize [inverse of (2.27)] or stabilize through radiative decay [(2.28)], the overall cross-section for dielectronic recombination can be written

$$\sigma_{DR} = \sigma_c \frac{A(R)}{A(R) + A(auto)}, \tag{2.29}$$

where σ_c, the capture cross-section, applies to reaction (2.27) and where the so-called *branching ratio* $A(R)/(A(R) + A(auto))$ contains the decay probabilities involved, namely the radiative rate $A(R)$ and the auto-ionizing rate $A(auto)$. The capture cross-section is often approximated by the excitation cross-section as given by van Regemorter (1962); see also Mewe (1972) and Alam and Ansari (1985). Also, one may assume detailed balance between capture and auto-ionization, corresponding to an equilibrium electron temperature, T_e, and put $\sigma_c = \text{const } A(auto)$. In this case, equation (2.29) takes the form

$$\sigma_{DR} = f(T_e, E^{**}) \frac{A(R) A(auto)}{A(R) + A(auto)}, \tag{2.30}$$

where $f(T_e, E^{**})$ is a function of the temperature and the energy of the intermediate state $N^{+(Z-1)**}$ and contains statistical weights of the ions. We shall see in Chapter 8 that dielectronic satellite lines of calcium and iron can provide important diagnostics of conditions in the hot coronal plasma during a flare.

2.3 Line profiles

According to equation (2.24), when an electron jumps from level j to level i, we should expect an infinitely narrow spectral line with frequency v_{ji} if the energy levels E_j and E_i are sharply defined. According to Heisenberg's uncertainty principle this would correspond to an infinitely long lifetime of the electron in that energy level. In reality, however, the lifetime is finite and, therefore, so is the width of the energy level. This is expressed quantum mechanically by Weisskopf and Wigner's (1930) probability distribution law which shows that the energy of a level with mean energy E_j is distributed as follows

$$P(E_j)\,dE = \frac{A_j}{h}\frac{dE}{\frac{1}{4}A_j{}^2 + \frac{4\pi^2}{h^2}(E - E_j)^2}. \tag{2.31}$$

Expression (2.31) defines a line profile whose shape we refer to as a *damping profile* or a *dispersion curve*; see curve D in Figure 2.3.

A_j is the probability of transitions from level j to all other possible levels. The probability function $P(E_j)$ has a sharp maximum for $E = E_j$, so that most of the atoms have energies near the mean energy E_j. When an electron jumps from level j to level i, the intensity of the ensuing line will depend on the widths of both levels and is then given by

$$I_{ji} = h\nu_{ji}A_{ji}\frac{A_j + A_i}{\frac{1}{4}(A_j + A_i)^2 + \frac{4\pi^2}{h^2}(E - (E_i - E_j))^2}, \tag{2.32}$$

where A_{ji} is the probability for the transition $j \to i$. This equation can be written

$$I = \frac{\text{const}}{\frac{\gamma_R{}^2}{4\pi} + (\nu - \nu_{ji})^2}, \tag{2.33}$$

where $\gamma_R = A_j + A_i$ is the radiation damping constant and is the sum of the

Fig. 2.3. Line profiles: G = Gaussian shape, D = damping profile, $\Delta\lambda$ = full width at half-intensity.

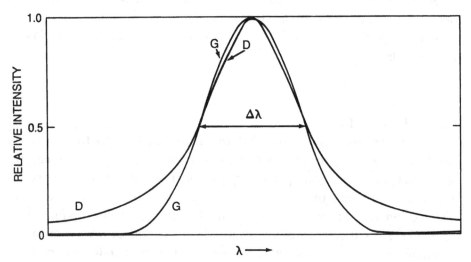

decay constants for the two energy levels involved in electron jump. The curve, sometimes called a Lorentzian, gives the so-called natural half-width of the line, which is a fundamentally lower limit to its width.

In the classical treatment of radiation where the atom is viewed with an oscillating electron executing forced vibrations, v_0, under the influence of the electric vector of the radiation field, one finds an expression for the intensity formally equal to equation (2.33) and with the damping constant given by

$$\gamma_{cl} = \frac{8\pi^2 e^2 v_0^{\,2}}{3m_e c^3}. \tag{2.34}$$

Expressing γ_{cl} in angstroms, we find $\gamma_{cl} = 1.8 \times 10^{-4}$ which means that radiation damping always constitutes an insignificant fraction of the total broadening of any solar line. The quantity γ_R also gives the half-width of the absorption coefficient, another concept of great importance in spectroscopy. The line absorption coefficient per atom measures the absorption of incident radiation of frequency v in a given direction. The total absorption is then found by integrating over all directions. We define the probability that a photon of frequency v will be absorbed by an atom as $(4\pi/hv)\alpha_v$ and find

$$\alpha_v = \frac{\pi e^2}{mc} \frac{\gamma_R}{4\pi^2} \frac{f}{(v - v_0)^2 + \dfrac{(\gamma_R)^2}{4\pi}}, \tag{2.35}$$

where f is the *oscillator strength* (or Landenberg f-value) and gives the number of equivalent classical oscillators, or the effective number of electrons in the atom. The integral over the absorption profile yields

$$\alpha = \int_0^\infty \alpha_v \, dv = \frac{\pi e^2}{mc^2} f.$$

The important case where the broadening of a spectral line is entirely due to the thermal motions of the atoms (or ions) is often found in flare physics. The width of such a line is given by the Doppler width $\Delta\lambda_D$

$$\Delta\lambda_D = \frac{\lambda_0}{c} \left(\frac{2kT}{m} \right)^{1/2}, \tag{2.36}$$

which results from combining the wavelength shift (or frequency shift) $\Delta\lambda_D$ due to the motion v_0 in the line of sight ($\Delta\lambda_D/\lambda_0 = v_0/c$) with the expression for v_0 in terms of the temperature

$$v_0 = \left(\frac{2kT}{m} \right)^{1/2}.$$

λ_0 is the wavelength of the undisturbed line center.

If the atoms in addition to their thermal velocity also possess a nonthermal (turbulent) component, ξ, then we can express the Doppler width by

$$\Delta\lambda_{\rm D} = \frac{\lambda_0}{c}\left(\frac{2kT}{m} + \xi^2\right)^{1/2}. \tag{2.37}$$

2.4   The temperature concept

We noticed above that under conditions of LTE one needs only one temperature parameter, and the source function is equal to the Planck function at that temperature [see equation (2.4)]. This temperature is the *kinetic temperature* $T_{\rm k}$, $S_\nu = B_\nu(T_{\rm k})$. The Boltzmann equation relating the population of two atomic levels gives this ratio in terms of an *excitation temperature* $T_{\rm ex}$

$$\frac{n_j}{n_i} = \frac{g_j}{g_i}\exp\left(-\frac{\chi_{ji}}{kT_{\rm ex}}\right). \tag{2.38}$$

When LTE prevails $T_{\rm k} \equiv T_{\rm ex}$, but in the low-density, high-temperature plasma of the solar atmosphere and under flare conditions, this identity does not hold. Under such conditions Thomas (1957) showed that the source function can be written

$$S_\nu = B_\nu(T_{\rm ex})\frac{\psi_\nu}{\phi_\nu} = \frac{2h\nu^3}{c^2}\left(\frac{n_i g_j}{n_j g_i} - 1\right)^{-1}\frac{\psi_\nu}{\phi_\nu}, \tag{2.39}$$

a form we introduced as equation (2.6). In general, ψ_ν/ϕ_ν varies with frequency across the line and with the viewing angle (between the direction to the Sun's center and the line-of-sight). In that case it is meaningless to speak of one excitation temperature at a point. Only if $\psi_\nu = \phi_\nu$ does $T_{\rm ex}$ have its usual meaning of the excitation temperature, and the source function reduces to

$$S_\nu = B_\nu(T_{\rm ex}). \tag{2.40}$$

A useful expression for the central intensity, I_0, of an emission line can be derived if we write the intensity of the line as

$$I_\nu = \int S_\nu e^{-\tau_\nu}\,{\rm d}\tau_\nu. \tag{2.41}$$

Equation (2.41) holds particularly for limb observations where the chromospheric contribution is negligible. Combining (2.40) and (2.41) we find for a line of small optical depth τ_0

$$I_0 = \tau_0 B_\nu(T_{\rm ex}).$$

We can therefore define a *radiation temperature* T_R by

$$B_\nu(T_R) = \tau_0 B_\nu(T_{ex}).$$
(2.42)

The same equation is often used to define a radiation temperature for the continuum. Equation (2.42) shows that when the optical depth is small, the radiation temperature is significantly smaller than the excitation temperature.

2.5 Line broadening mechanisms

The shape of a spectral line is often used to assess the temperature or the density of the radiating plasma. In certain cases, the width of a line is uniquely given by the temperature [see equation (2.36)], but often other factors play a role in broadening the line. Sometimes nonthermal velocities are involved, e.g., equation (2.37); in other instances, the high density of a plasma results in an excess broadening.

2.5.1 *Collisional damping*

In our discussion of spectroscopy thus far we have not considered any influence on the radiating atoms from their neighbors, and this may be a fair approximation in tenuous plasmas such as the quiet solar corona. However in flares, where the pressure can be appreciable, collisions of radiating atoms or ions with surrounding particles lead to a broadening of the line called *pressure broadening* or *collisional damping*.

The collision processes are very complicated, and a general theory is not available. Two limiting cases are normally considered, depending on the point of view used to treat the collision. The discrete encounter theory, or collision damping theory, which goes back to Lorentz (1905*a, b*) [see, in particular, Lindholm (1942)], considers the radiating atom as disturbed by separate, discrete collisions, whereby the wave emitted (or absorbed) by the atom is interrupted. The resulting line intensity is computed by a Fourier analysis of the wave train.

The other approach, the statistical theory, goes back to Stark (1916), Holtsmark (1919), and others, and Stark broadening and Holtsmark's theory are still current concepts. The method assesses the value – at a given atom or ion – of the field produced by the surrounding particles. Each field causes a shift in the frequency of the radiation from the atom, and by assigning statistical weights to the shifts, the line profile is found by integrating over all shifts.

The two approaches are used for different atoms and for different circumstances. Often there is a critical distance, $\Delta\lambda_{crit}$, measured from the line center, within which the broadening may be described by the discrete encounter theory, while for distances greater than $\pm\Delta\lambda_{\text{crit}}$, the profile is determined by the statistical theory.

Let a neighboring particle cause a change δv in the frequency of a radiating atom. In the statistical theory one analyzes these changes as a function of the distance between the two atoms and assumes that the functional form is

$$\delta_v = \frac{C}{r^s},$$

where s is an integer; C is constant. If $s = 2$ the statistical broadening is symmetrical, but for $s = 4$ or $s = 6$ (van der Waals forces) the broadening is asymmetrical and the line is both broadened and shifted.

One of the best known examples of statistical broadening is the *Stark effect* where the electrostatic field is due to the charged particles in the plasma. The spectral line is split into several components in the electric field, E, and a given component i is displaced by an amount Δv_i which depends on the field. In some instances, such as hydrogen, Δv_i is directly proportional to E; i.e.,

$$\Delta v_i = \frac{3h}{8\pi^2 me} q_i E. \tag{2.43}$$

This is the *linear* Stark effect; q_i is a number, which for the Balmer lines is given by $q \leqslant n(n-1) + 1$, n being the number of the Balmer line (e.g., H_{20}).

In other atoms and ions Δv_i is proportional to the square of E, i.e.,

$$\Delta v_i = C_i E^2. \tag{2.44}$$

This is the *quadratic* Stark effect.

The influence of the Stark effect on hydrogen lines in solar flares has been extensively investigated (see Švestka, 1965, 1976, and references therein), and the wings of the higher Balmer lines furnish a good diagnostic tool for density determinations. As the density of the plasma increases, the wings of the lines become more and more pronounced. Canfield *et al.* (1984) have carried out an extensive analysis of this wing broadening effect on the $H\alpha$ line in order to test models of energy transport during solar flares (see Chapter 7). Further, since the Balmer lines lie closer and closer together, the higher their number, the more the wings of neighboring lines will overlap as a result of this broadening. In other words, the higher the density, the fewer lines that can be resolved. This situation was used by Inglis and Teller (1939) and later by Kurochka and Maslennikova (1970), who derived a relationship between the electron density and the quantum number, n_{max}, of the highest resolvable Balmer line:

$$\log n_e = 22.7 - 7.0 \log n_{max}. \tag{2.45}$$

If for example one can resolve up to $n_{max} = 30$ in a certain spectrum, equation (2.45) gives a value of $10^{12}\,\mathrm{cm}^{-3}$ for the electron density.

2.5.2 *Doppler broadening*

In addition to pressure broadening the most important broadening mechanism we will encounter in flares is Doppler broadening. It is due to the fact that the photons we observe are not radiated from atoms at rest with respect to the observer. At any finite temperature, the radiating atoms move with a range of velocities which depends on the value of the temperature. Let the velocity component of the radiating atom along the line of sight be v and the frequency of radiated photons (i.e., in the atom's frame) be v_0. The observed frequency will be $v_0' = v_0[1 - (v/c)]$ and the expression for the absorption coefficient (2.35) takes the form

$$\alpha_v' = \frac{\pi e^2}{mc}\frac{\gamma_R}{4\pi^2}\frac{f}{\left(v - v_0 + \dfrac{vv_0}{c}\right)^2 + \left(\dfrac{\gamma_R}{4\pi}\right)^2}. \tag{2.46}$$

To find the average absorption coefficient we must integrate over the velocity distribution $f(v)$. If the latter is Maxwellian, we find

$$\alpha_v = \int_{-\infty}^{+\infty} \alpha_v' f(v)\, dv$$

$$= \frac{\pi e^2}{(\pi mc)^{1/2}}\frac{\gamma_R}{4\pi^2} f \int_{-\infty}^{+\infty} \frac{\exp\left[-\left(\dfrac{\Delta v^2}{\Delta v_0}\right)\right] d(\Delta v/\Delta v_D)}{\left(v - v_0 + \dfrac{vv_0}{c}\right)^2 + \left(\dfrac{\gamma_R}{4\pi}\right)^2}, \tag{2.47}$$

where $\Delta v_D/v_0 = v_0/c$ and $v_0 = (2kT/m)^{1/2}$. This is known as a *Voigt profile*.

Since (2.47) cannot be evaluated in closed form, we shall look at two limiting cases by writing the expression in the form

$$\alpha_v = \mathrm{const} \int_{-\infty}^{\infty} \frac{\exp(-X^2)\, dx}{(X - Y)^2 + Z^2}, \tag{2.48}$$

where

$$X = \frac{\Delta v}{\Delta v_0} = \frac{v}{(2kT/m)^{1/2}}, \quad Y = \frac{v_0 - v}{v_0}\frac{c}{v_0}, \quad \text{and} \quad Z = \frac{\gamma_R \dfrac{c}{4\pi}}{vv_0}.$$

In the first limiting case $Y \gg X$. We then retrieve the original form of the absorption coefficient (2.46), which means that the thermal motions have no influence on the absorption of the radiation. This case applies to the far wings of the line. In the other limiting case both Y and Z are small and the profile reduces to

$$\alpha_v = \text{const} \frac{4\pi}{\gamma_R} \exp(-Y^2) \quad \text{or} \quad \alpha_v \propto \exp\left[-\left(\frac{v-v_0}{v_0}\right)^2 \frac{c^2}{v_0^2}\right], \quad (2.49)$$

which shows the dominating influence of the Maxwellian distribution of the radiating atoms, while the original damping shape of the line has completely disappeared; see curve G in Figure 2.3.

2.6   The differential emission measure

With the advent of space-borne spectrographs, observations of spectral lines in soft X-rays (approximately 1–100 Å), the extreme ultraviolet (EUV approximately 100–1000 Å), and the moderate ultraviolet (1000–2500 Å) have made possible new diagnostics for temperature – and density – determinations of the flare plasma. The basic physics involved is the same as treated in Section 2.2, but it has been necessary to work out the atomic data for many more ions, e.g., Ca XIX and Fe XXI which at $\approx 20 \times 10^6$ K give lines around 3.2 Å and 1.8 Å, respectively, Mg X which at 10^6 K produces the 625 Å line, or O V and C IV which at 80–100 000 K are responsible for the 1371 Å and 1548 Å lines respectively. By studying profiles of and ratios between selected lines from these and other ions, temperature and density of the plasma emitting the lines can be deduced. We will return to specific examples in Chapter 8. For details of the X-ray, EUV, and UV diagnostics involved we refer the reader to excellent reviews by Feldman (1980) and Doschek (1985).

To draw conclusions from spectral observations, we often employ the differential emission measure, $Q(T)$. This parameter can be derived from spectral observation with a minimum of assumptions (Withbroe, 1977). It is defined by

$$\int Q(T) \, dT = \int n_e^2 \, dV \qquad (2.50)$$

and gives the emitting power at temperatures between T and $T + dT$ contained in the volume V [see Craig and Brown (1976) for a more rigorous definition]. We can then use Q to express the total flux F_l emitted in a spectral line. An expression for F_l which is valid for a two-level atom (Pottasch, 1964; Withbroe, 1970) is

$$F_l = 2.2 \times 10^{-15} Af \int gn_e^2 \cdot G(T) \, dV, \qquad (2.51)$$

where A is the chemical abundance relative to hydrogen, f the oscillator strength, g the Gaunt factor, which takes into account quantum mechanical effects absent in the classical description (Gaunt, 1930; van Regemorter, 1962), and $G(T)$ a temperature-dependent term relating to the excitation and ionization properties of the atom or ion producing the spectral line. The density appears raised to the second power because of the dominant collisional excitation process in which the radiated power per atom is proportional to n_e [equation (2.16)]. Note that this is a macroscopic way of describing the emission in a spectral line that we have previously treated in a microscopic way, i.e., by equation (2.22) for a line between levels i and j. For atoms more complicated than the two-level model atom, the density and temperature dependences cannot be separated and expressed so conveniently as a product $n_e^2 \cdot G(T)$. Combining equations (2.50) and (2.51) we find

$$F_l = 2.2 \times 10^{-15} A f \int Q(T) g G(T) \, dT. \tag{2.52}$$

To calculate the function $G(T)$ one assumes ionization equilibrium which should be valid during the gradual phase of flares (Chapter 8) when relaxation of the plasma sets in after the impulsive phase (Chapter 7).

In practice one can use spectral lines in the UV region [e.g., the Mg X, 625 Å or O VI, 1032 Å lines (Reeves *et al.*, 1977) or the O V, 1371 Å line (Tandberg-Hanssen *et al.*, 1980)] to get data on plasmas at temperatures of the order 10^6 K or less, and then use X-ray data [e.g., lines of Ca XIX around 3.2 Å (Gabriel *et al.*, 1981)] to obtain information about the plasma in excess of a million degrees. By analogy with equation (2.52), the X-ray flux can be written

$$F_X = \int Q(T) G_X(T) \, dT, \tag{2.53}$$

where the function $G_X(T)$ is given by

$$G_X(T) = \frac{F_X}{\int n_e^2 \, dV}. \tag{2.54}$$

Dere *et al.* (1974) have calculated $G_X(T)$ (as a function of T) for the three wavelength bands 0.5–3 Å, 1–8 Å, and 8–20 Å observed by Solrad 9, and Withbroe (1975) has devised an iterative method whereby one can estimate $Q(T)$ from measurements of F_l and F_X, using equations (2.52) and (2.53). We note (Craig and Brown, 1976) that the determination of $Q(T)$ is neither well posed nor unique – only its integral over temperature is well defined observationally, and many different $Q(T)$ can have the same $\int Q(T)\,dT$. In practice, therefore, certain assumptions (e.g., smoothness, monotonicity) are employed in the reduction process.

2.7   Influence of magnetic fields

In Section 2.2.1 we remarked that the presence of a magnetic field in a radiating plasma will remove the degeneracy of atomic levels, which then are split into their M_J sublevels or states. Under such conditions the spectral lines emitted – or absorbed – by the atom will be changed, and by observing these spectral line changes, one can deduce the properties of the magnetic field. These changes can be described either semi-classically or quantum mechanically, and are due to what we call the *Hanle* and *Zeeman* effects. In these cases we deal with polarized radiation, and we shall start this section by a brief description of polarized photons.

2.7.1   *Description of polarized radiation*

We can define polarization of electromagnetic waves in terms of the electric field vector in planes perpendicular to the direction of propagation. If the radiation is unpolarized, the electric vector is randomly oriented; otherwise, it will either be always oriented in the same direction, i.e., linearly polarized waves, or the projection of the tip of the vector will describe an ellipse, i.e., elliptically polarized waves (of which circularly polarized waves form a subset). Let a and b be the semi-major and semi-minor axes, respectively, of the polarization ellipse, and I_a and I_b the intensity of the radiation with the electric vector along a and b. We define the percentage polarization as

$$p = 100\frac{I_a - I_b}{I_a + I_b}. \tag{2.55}$$

The ellipticity of the polarized ellipse is $\sin \chi$, where $\tan \chi = b/a$ [e.g., Born and Wolf (1965)].

The description of polarization can be completely and elegantly cast in the framework of the Stokes parameters [see Stokes (1852); Perrin (1942)]. We consider the radiation propagating in the Z-direction as consisting of two orthogonal plane waves, $E_x(t)$ and $E_y(t)$, viz.,

$$E_x(t) = E_{0x} \exp[i(\omega t + \delta_x)] = \varepsilon_{0x} \exp(i\omega t) \tag{2.56}$$

and

$$E_y(t) = E_{0y} \exp[i(\omega t + \delta_y)] = \varepsilon_{0y} \exp(i\omega t), \tag{2.57}$$

where ε_{0x} and ε_{0y} are complex amplitudes containing the phase factors δ_x

and δ_y. The Stokes parameters are conveniently written as a column matrix

$$
S = \begin{Bmatrix} I \\ Q \\ U \\ V \end{Bmatrix} = \begin{Bmatrix} \varepsilon_{0x}\varepsilon_{0x}{}^* + \varepsilon_{0y}\varepsilon_{0y}{}^* \\ \varepsilon_{0x}\varepsilon_{0x}{}^* - \varepsilon_{0y}\varepsilon_{0y}{}^* \\ \varepsilon_{0x}\varepsilon_{0y}{}^* + \varepsilon_{0y}{}^*\varepsilon_{0y} \\ \varepsilon_{0x}\varepsilon_{0y}{}^* - \varepsilon_{0x}{}^*\varepsilon_{0y} \end{Bmatrix} = \begin{Bmatrix} E_{0x}{}^2 + E_{0y}{}^2 \\ E_{0x}{}^2 - E_{0y}{}^2 \\ 2E_{0x}E_{0y}\cos(\delta_y - \delta_x) \\ 2E_{0x}E_{0y}\sin(\delta_y - \delta_x) \end{Bmatrix}. \quad (2.58)
$$

The last part of equation (2.58) indicates the relationship of the Stokes four-vector representation to the description using the polarization ellipse, since the ellipticity can be written

$$
\sin \chi = \frac{V}{(Q^2 + U^2 + V^2)^{1/2}}.
$$

Furthermore,

$$
I^2 \geqslant Q^2 + U^2 + V^2, \quad (2.59)
$$

where the equality sign pertains to completely polarized light.

The parameter I is the total intensity of the radiation, Q and U describe the linear, and V the elliptical state of polarization. The Stokes parameters are therefore the observables of the polarization ellipse, or of the beam of radiation. For incompletely polarized light, i.e., when (2.59) is a strict inequality, the intensity just referred to represents the polarized part, I_p, only, so that the degree of polarization is

$$
p = \frac{I_p}{I} = \frac{(Q^2 + U^2 + V^2)^{1/2}}{I};
$$

cf. (2.55).

2.7.2 *The Zeeman effect*

When a magnetic field removes the degeneracy of an atomic level, we observe – instead of the single emission line encountered in the absence of a magnetic field – several line components, each due to a transition from one of the split states of the upper level. This condition is referred to as the *emission Zeeman effect*. Since the splitting of the level follows rules for M_J described in 2.2.1, the Zeeman components of a line are located in the spectrum symmetrically about the position λ_0 of the undisturbed line. An atom (or ion) in a magnetic field can also absorb photons by having its valence electron raised to any of the split states of the upper level in question. This means that a normally undisturbed absorption line will be split into several components in a magnetic field.

The simplest case to consider is that of a line of a singlet series, where the splitting is the so-called normal triplet, having wavelengths $\lambda_0 - \Delta\lambda_B$, λ_0, and $\lambda_0 + \Delta\lambda_B$. The undisplaced line is called a π-component, the two displaced lines σ-components. In the transverse case, i.e., observing perpendicularly to the direction of the magnetic field, the π-component is plane polarized parallel to the field for an emission line and perpendicular to the field for an absorption line. In this transverse case, the σ-components (at $\lambda_0 \pm \Delta\lambda_B$) are plane polarized perpendicular to the field for an emission line and partially plane polarized parallel to the field in an absorption line [e.g., Bray and Loughhead (1965)]. Turning now to the longitudinal case, i.e., observing along the magnetic field, we notice that the π-component (at λ_0) is absent. Further, with the magnetic field pointing toward the observer, we find that the $\sigma(\lambda_0 - \Delta\lambda_B)$ component is left-handed circularly polarized and the $\sigma(\lambda_0 + \Delta\lambda_B)$ component is right-handed circularly polarized for an emission line, and vice versa for an absorption line. The polarizations of the σ-components are reversed for a magnetic field directed away from the observer.

The wavelength shift $\Delta\lambda_B$ of the σ-components is given by

$$|\Delta\lambda_B| = \frac{e}{4\pi m_e c^2}\lambda_0{}^2 g B = 4.67 \times 10^{-13}\lambda_0{}^2 g B, \tag{2.60}$$

where λ_0 is in angstroms and the magnetic field B is in gauss. If we work in frequency units, the line shift is given by $\Delta\nu_B = eBg/4\pi m_e c$. The Landé g-factor determines the amount of splitting of the line in terms of the magnetic moment, μ_B, of the atom or ion in the direction of B,

$$\mu_B = \frac{eh}{4\pi m_e c}gM, \tag{2.61}$$

where M is the magnetic quantum number and follows the rules [equation (2.25)]; i.e., $\Delta M = \pm 1$. When $\Delta M = 0$ we observe the unshifted π-component. If Russell–Saunders coupling prevails, Landé's g-factor is a simple function of the S, L, and J quantum numbers and can easily be determined from the expression (Condon and Shortley, 1953)

$$g = 1 + \frac{J(J + 1) - L(L + 1) + S(S + 1)}{2J(J + 1)}. \tag{2.62}$$

For strong magnetic fields, the two sigma components are well separated and the field can be determined from equation (2.60) by measuring the displacement $\Delta\lambda_B$. However, since this procedure requires fields of several thousand gauss, the method lends itself only for observations of sunspot fields. Otherwise one makes use of the polarization properties of the Zeeman effect (Thiessen, 1949; Babcock and Babcock, 1953, 1955).

In equation (2.58) we wrote the Stokes parameters as the four elements of a single-column matrix $\{I, Q, U, V\}$. For an absorption line, it is possible to write the intensities of the Zeeman components in terms of the absorption coefficients κ [e.g., Bray and Loughhead, 1965)] and in terms of the emission coefficients ε for an emission line, viz.

$$
\left.
\begin{aligned}
I_\pi &= \varepsilon_\pi \{\sin^2 \gamma, \sin^2 \gamma, 0, 0\}, \\
I_{\sigma,r} &= \varepsilon_{\sigma,r} \{\tfrac{1}{2}(1 + \cos^2 \gamma), -\tfrac{1}{2}\sin^2 \gamma, 0, \cos \gamma\}, \\
I_{\sigma,l} &= \varepsilon_{\sigma,l} \{\tfrac{1}{2}(1 + \cos^2 \gamma), -\tfrac{1}{2}\sin^2 \gamma, 0, -\cos \gamma\},
\end{aligned}
\right\}
\tag{2.63}
$$

where r and l denote right and left circular polarizations, respectively, γ is the angle between the line of sight and the direction of the magnetic field, and the emission coefficients have the form

$$
\left.
\begin{aligned}
&\varepsilon_\pi = \varepsilon(\nu_0), \quad \text{i.e., identical to the coefficient in the} \\
&\qquad \text{absence of a field}, \\
&\varepsilon_{\sigma,r} = \varepsilon(\nu_0 + \Delta\nu_B) = \varepsilon_\nu + \Delta\nu_B \frac{\partial \varepsilon}{\partial \nu} + \cdots \\
&\qquad\qquad\qquad\qquad \text{(Taylor series for small } B), \\
&\varepsilon_{\sigma,l} = \varepsilon(\nu_0 - \Delta\nu_B) = \varepsilon_\nu - \Delta\nu_B \frac{\partial \varepsilon}{\partial \nu} + \cdots.
\end{aligned}
\right\}
\tag{2.64}
$$

The case of absorption lines is referred to as the *inverse Zeeman effect*. The general treatment of radiation transfer in these cases is very complicated. We obtain three equations corresponding to equation (2.3) in the field-free case. Unno (1956) derived the equations of transfer under the assumption that the line is formed by true absorption. They have the form [cf. Chandrasekhar (1950); see also Stepanov (1958)]

$$
\left.
\begin{aligned}
\mu \frac{dI_\nu}{d\tau_\nu} &= (1 + \eta_I)I_\nu + \eta_Q Q_\nu + \eta_V V_\nu - (1 + \eta_I)S_\nu, \\
\mu \frac{dQ_\nu}{d\tau_\nu} &= \eta_Q I_\nu + (1 + \eta_I)Q_\nu - \eta_Q S_\nu, \\
\mu \frac{dV_\nu}{d\tau_\nu} &= \eta_V I_\nu + (1 + \eta_I) V_\nu - \eta_V S_\nu,
\end{aligned}
\right\}
\tag{2.65}
$$

where

$$
\left.
\begin{aligned}
\eta_I &= \frac{1}{2\kappa_\nu} \left[\kappa_\pi \sin^2 \gamma + \tfrac{1}{2}(\kappa_{\sigma,r} + \kappa_{\sigma,l})(1 + \cos^2 \gamma) \right], \\
\eta_Q &= \frac{1}{2\kappa_\nu} \left[\kappa_\pi - \tfrac{1}{2}(\kappa_{\sigma,r} + \kappa_{\sigma,l}) \right] \sin^2 \gamma, \\
\eta_V &= \frac{1}{2\kappa_\nu} (\kappa_{\sigma,r} - \kappa_{\sigma,l}) \cos \gamma.
\end{aligned}
\right\}
\tag{2.66}
$$

Analytic solutions for equation (2.65) are possible only if further simplifying assumptions are made, viz. uniform magnetic field and Milne–Eddington model atmosphere (κ_π/κ_v, $\kappa_{\sigma,\mathrm{r}}/\kappa_v$, and $\kappa_{\sigma,\mathrm{l}}/\kappa_v$ do not vary with depth).

2.7.3   *The Hanle effect*

More than 60 years ago it was discovered (Strutt, 1922) that if one excites a mercury gas with polarized radiation, the emitted resonance line (at 2537 Å) is linearly polarized. This resonance polarization is a special case of fluorescence polarization. In general, emission lines will exhibit fluorescence polarization when the incident photon flux that excites the upper level of the transition in question is either polarized or anisotropic. Similarly, what is called impact polarization is caused by excitation with a flux of charged particles. Resonance polarization will be affected by the presence of a magnetic field in the radiating plasma, an effect studied by Hanle (1924, 1925) and referred to as the Hanle effect.

The resulting polarization, both its direction and amount, depends on the direction of polarization of the exciting radiation as well as on both the direction and strength of the magnetic field. Sometimes one observes strong polarization when an essentially unpolarized line appears in the field-free case; while, in other instances, the magnetic field depolarizes the line. In the simple case of mercury vapor, the Hanle effect can be qualitatively understood by the classical theory of the atom; but, in other cases, a quantum mechanical treatment is required.

In the classical theory, the scattering atom behaves like a classical oscillator. If the direction of the vibration of the electric vector in the exciting radiation is parallel to the X-axis and there is a magnetic field (25–100 gauss) in the same direction, the polarization of the resonance-scattered light will be in the same direction as in the field-free case; i.e., we see that a high degree of incident polarization along the X-axis produces scattered light that is highly polarized in the X-direction. This follows immediately from classical theory of the atom in which the scatterer behaves like a classical oscillator. However, if the (weak) magnetic field is parallel to the Y-axis, we observe a strong resonance-scattered line which is highly polarized in the X-direction. If the field strength increases, one observes that the percentage polarization decreases. So long as the field is weak, the direction of polarization deviates only slightly from that of the field-free case, but as the field increases – and the polarization decreases – the plane of polarization rotates through large angles.

This sensitivity of the observed polarization to the orientation of the magnetic field is explained by invoking the precession of the atom about the magnetic vector, and the sensitivity of polarization to the strength of the field is explained by assuming that the atom is a damped oscillator. Due to the precession, the direction of vibration will deviate from the original direction and cause the plane of polarization of the emitted light to rotate. The oscillator describes a rosette (see Figure 2.4) when viewed along the magnetic field, and the shape of the rosette – and therefore the nature of polarization – will depend on the ratio between the angular velocity of precession, ω, and the radiative damping constant, γ_R (or the reciprocal of the mean radiative lifetime, τ), of the oscillator. If $\omega \gg \gamma_R$, or

$$\omega\tau \gg 1, \tag{2.67}$$

Fig. 2.4. Path described by an oscillator when the angular velocity of precession, ω, is of the order of the radiative damping constant $1/\tau$.

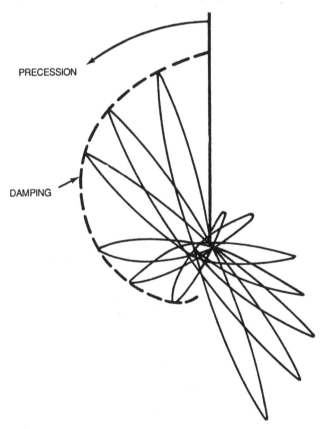

PRECESSION

DAMPING

$\omega\tau \lesssim 1$

the rosette will be axially symmetric, since the atom will have ample time to precess before it is damped out. Consequently, there is no polarization observed along the magnetic field if the field is strong enough to make inequality (2.67) hold. If $\omega \approx \gamma_R$, or $\omega\tau \approx 1$, the oscillator describes an asymmetrical rosette which means that the degree of polarization is reduced relative to the value in the field-free case, and the direction of polarization is rotated an angle Ω with respect to the direction of polarization of the exciting beam. Finally, if $\omega\tau \ll 1$, the oscillator hardly has time to precess before it is damped out, which explains the weak field results.

Breit (1925) derived expressions for the percentage polarization, P, and the angle of rotation, Ω. For polarization observed along the magnetic field of strength B, the polarization is

$$p = \frac{p_0}{1 + (g\Omega_B\tau)^2},\qquad(2.68)$$

where p_0 is the percentage polarization for the field-free case and $\Omega_B = eB/2m_e$ is the Larmor precession velocity for an electron of mass m_e and charge e. The angle of rotation is given by

$$\tan 2\Omega = g\Omega_B\tau.\qquad(2.69)$$

In equations (2.68) and (2.69), g is a correction factor necessary to obtain the observed results from classical theory, and Breit (1933) showed that classical and quantum mechanical results agree if this correction factor is Landé's g-factor for the upper level of the transition in question; equation (2.62).

An elementary quantum mechanical treatment, involving the Zeeman states of the levels for the transition in question, was first used by Hanle (1923). The behavior of the mercury resonance line polarization can be understood from the appropriate energy level diagram for the $^1S_0-^3P_1$ transition responsible for the 2537 Å line; see Figure 2.5. If the $^3P_1$ level is excited by light polarized parallel to the X-axis and if there is a strong magnetic field in the same direction, only the π-component (which, according to the classical Zeeman effect, is polarized parallel to the field) can be absorbed. Hence, we populate only the $M_J = 0$ Zeeman state of the $^3P_1$ level. Then, if there is no inter-level transition taking place, the emitted radiation will be a π-component whose electric vector is parallel to the field. On the other hand, if the magnetic field is parallel to the Z-axis, the circularly polarized components will be absorbed. The circularly polarized components will subsequently be re-emitted, but along the line-of-sight, the Y-axis, which is perpendicular to the field, the radiation will appear linearly polarized parallel to the X-axis.

In the field-free case and when the upper level is degenerate, one cannot *a priori* predict which components will be absorbed. This difficulty is resolved by Heisenberg's (1926) principle of spectroscopic stability, which states that the absorption of radiation in the field-free case is the same as when the magnetic field is parallel to the direction of vibration of the incident polarized beam; i.e., one assumes that absorption occurs in π-components.

When fluorescence polarization is analyzed using simple scattering theory in a normal Zeeman pattern, we can account for the observed phenomena only when dealing with simple cases of resonance radiation. In more general cases where scattering takes place between states only weakly removed from degeneracy by a magnetic field, mutual interactions between close atomic levels, i.e., coherence effects, come into play. This shows that the Hanle effect should be understood as a coherence effect, already pointed out by Warwick and Hyder (1965). The first attempt to include coherence arguments was made by Obridko (1968) in a phenomenological description. General

Fig. 2.5. σ and π Zeeman components due to transitions $^1S_0 - {}^3P_1$.

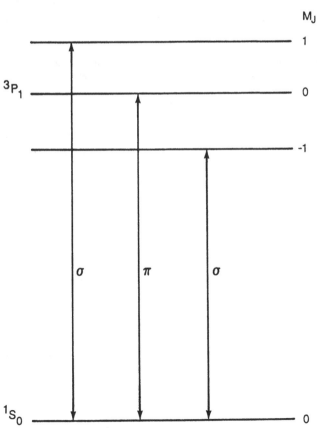

discussions of the coherence problem have been given by House (1970a, b; 1971), and applications to solar physics have been considered by Leroy *et al.* (1977) and Sahal-Bréchot *et al.* (1977), Landi Degl'Innocenti (1982) and others.

Referring to Figure 2.6, we consider a simple, and more complex, case of scattering. In Figure 2.6(*a*) the sublevels are far removed from degeneracy, so that each M_J state can be treated independently of the others. If we neglect collisions and interlocking with other levels, we see that if an excitation raises an electron from the M_J state of level *a* to the excited state M_J'' of level *b*, the subsequent emission of radiation will be due to a transition from this state M_J'' back to a state M_J' of level *a*. However, if we have a situation where the states are not far removed from degeneracy, Figure 2.6(*b*), the different M_J'' states are more or less indistinguishable, and one must sum over those states that could be excited by the incident radiation. As pointed out by House (1971), the summation over the excited states is carried out before one squares the amplitudes to obtain the scattered intensity; and, from a mathematical point of view, it is the cross-terms that may result from this procedure that produce the coherence between the exciting radiation and the fluorescence emission that contains the physics behind the Hanle effect.

Fig. 2.6. Energy-level diagram showing quantum numbers and transitions where coherency effects are not encountered (*a*) and where such effects are important (*b*).

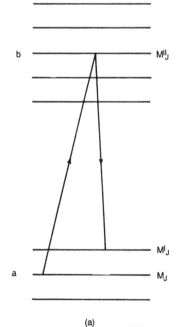

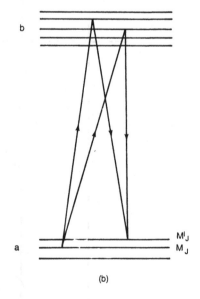

(a)

(b)

This completes our survey of the spectroscopic tools we shall need to interpret various observational data (magnetograph, soft X-ray, EUV, optical, etc.) pertaining to solar flares. We now turn to another fundamental tool – the magnetohydrodynamics of a plasma.

3

Elements of flare magnetohydrodynamics

All is flux, nothing stays still

Heraclitus (*c.* 540–*c.* 480 BC)

Magnetohydrodynamics constitutes the second important tool used in analyses of solar activity, particularly in the study of solar flares. While spectroscopy gives us means by which the plasma parameters, such as temperature and density, and plasma conditions, such as velocity and embedded magnetic fields, can be assessed, magnetohydrodynamics (MHD) furnishes the tool by which the interaction of motions and fields can be studied.

We have seen in Section 2.7 how the introduction of a magnetic field complicates the analysis of spectral lines, and we shall now discuss how the dynamics of a plasma, such as the Sun's atmosphere, is also affected – even controlled – by the presence of magnetic fields.

The quiet solar atmosphere is in constant motion, and in active regions, particularly in flares, we find that a dynamic description is even more essential for a proper understanding of the physics involved. Therefore, the governing equations will not be simply the conservation equations of gas or hydrodynamics. However, as we gradually include the effects of magnetic fields, let us first recall the framework of the gas dynamic conservation equations.

3.1   The Boltzmann equation and its moments

The equations that describe the state of a fluid are generally obtained from the Boltzmann equation for the velocity distribution function $f(\mathbf{r}, \mathbf{w}, t)$, i.e.,

$$\frac{\partial f}{\partial t} + \sum_j \frac{\partial f}{\partial x_j} \frac{\partial x_j}{\partial t} + \sum_j \frac{\partial f}{\partial w_j} \frac{\partial w_j}{\partial t} = \left(\frac{\partial f}{\partial t}\right)_{\text{coll}}, \quad j = 1, 2, 3$$

which may be written

$$\frac{\partial f}{\partial t} + \mathbf{w} \cdot \nabla_r f + \frac{\mathbf{F}}{m} \cdot \nabla_w f = \left(\frac{\partial f}{\partial t}\right)_{\text{coll}}, \tag{3.1}$$

where **F** is the external force experienced by the particles. If the fluid consists of more than one species of particle (such as an electron–ion plasma), each species will obey an equation of the form (3.1). The function $f(\mathbf{r}, \mathbf{w}, t)$ gives the density of particles of mass m in the six-dimensional phase-space $(\mathbf{r}, \mathbf{w})$ as a function of time. The term $(\partial f/\partial t)_{\text{coll}}$ is the collision integral, the term that makes the complete solution of the Boltzmann equation so complicated [see, e.g., Chapman and Cowling (1952)]. The operator ∇_r performs differentiation with respect to the space coordinates $\mathbf{r} = (x, y, z)$ or $= (x_1, x_2, x_3)$, while the operator ∇_w similarly indicates differentiation with respect to the velocity components $\mathbf{w} = (w_x, w_y, w_z)$ or $= (w_1, w_2, w_3)$. That is, we have

$$\mathbf{w} \cdot \nabla_r f = \sum_j w_j \frac{\partial f}{\partial x_j} \quad \text{and} \quad \frac{\mathbf{F}}{m} \cdot \nabla_w f = \sum_j \frac{F_j}{m} \frac{\partial f}{\partial w_j}, \quad j = 1, 2, 3$$

When there are no collisions $(\partial f/\partial t)_{\text{coll}} = 0$, and (3.1) reduces to the collisionless Boltzmann equation, or *Vlasov equation* [equation (4.84)], which holds approximately for a very tenuous plasma.

At this stage one may proceed in one or two directions. Either we can continue to treat the individual particles in a microscopic sense; that is, take recourse to a statistical theory where we seek information about the distribution functions for the particle velocities and introduce the Boltzmann equation, or we can take a macroscopic, 'fluid' point of view and develop the gas or hydrodynamic equations. The former, kinetic, framework should be used when, e.g., treating high-energy particle beams and their interaction with the solar atmosphere, and we will discuss some examples of this microscopic plasma physics approach in Chapter 4. However, there is a host of phenomena in active-Sun research where a continuous plasma dynamics viewpoint is adequate. These are large-scale phenomena where the particle mean free paths λ are short compared to the characteristic length L of the phenomenon under study, and the concept of a fluid temperature is valid. In the following we shall discuss this macroscopic approach and its application to a solar-type plasma where the total number density, n, is a sum of the electron density, n_e, and the ion density, n_i. With different ion species present, we have $n = n_e + \sum_s n_s$, with s representing the different ion species. For a simple two-component plasma, the total pressure is $p = p_e + \sum_s p_s \rightarrow p_e + p_i$, and for a perfect gas $p = k(n_e + n_i)T$, where the kinetic temperature of the plasma is defined as

$$T = \frac{1}{n}(n_e T_e + n_i T_i). \tag{3.2}$$

In the macroscopic picture we are interested not so much in the individual particle velocities **w** but rather in the flow velocity **v** given in terms of the mass density, $\rho = n_e m_e + n_i m_i \approx n_i m_i$, by the equation

$$\mathbf{v} = \frac{1}{M}\left(m_e \sum \mathbf{w}_e + m_i \sum \mathbf{w}_i\right) = \frac{1}{\rho}(n_e m_e \mathbf{v}_e + n_i m_i \mathbf{v}_i), \tag{3.3}$$

where M is the total mass of the plasma under consideration and

$$\mathbf{v}_i = \frac{1}{n_i \Delta V}\sum \mathbf{w}_i; \quad \mathbf{v}_e = \frac{1}{n_e \Delta V}\sum \mathbf{w}_e, \tag{3.4}$$

with ΔV being the plasma volume.

The motions of the electric charges on n_e and n_i lead to a current density

$$\mathbf{j} = e(n_i Z \mathbf{v}_i - n_e \mathbf{v}_e), \tag{3.5}$$

where Z is the number of elementary charges on the ions n_i. If we introduce the diffusion velocity **u**, which for the sth species is defined as $\mathbf{u}_s = \mathbf{v}_s - \mathbf{v}$, we can write the expression for the current density as

$$\begin{aligned}
j &= e(Zn_i u_i - n_e u_e) + ev(Zn_i - n_e) \\
&= \text{conduction current} + \text{convection current};
\end{aligned} \tag{3.5'}$$

the latter term constitutes the bulk flow of a charged fluid and vanishes for a neutral plasma.

We may proceed from the microscopic Boltzmann [equation (3.1)] to derive the macroscopic conservation equations by multiplying throughout by a suitable function of the microscopic velocity and then integrating over velocity space. Let this function be $Q(\mathbf{w})$. Then

$$\iiint_{-\infty}^{\infty} Q(\mathbf{w})\frac{\partial f}{\partial t}\,dw_x\,dw_y\,dw_z = \frac{\partial}{\partial t}\iiint_{-\infty}^{\infty} Q(\mathbf{w})f\,dw_x\,dw_y\,dw_z$$

$$= \frac{\partial}{\partial t}(n\bar{Q}), \tag{3.6}$$

where $\bar{Q}$ is the mean value of Q. Also,

$$\sum_j \iiint_{-\infty}^{\infty} Q(\mathbf{w})w_j\frac{\partial f}{\partial x_j}\,dw_x\,dw_y\,dw_z$$

$$= \sum_j \frac{\partial}{\partial x_j}\iiint_{-\infty}^{\infty} Q(\mathbf{w})w_j f\,dw_x\,dw_y\,dw_z$$

$$= \sum_j \frac{\partial}{\partial x_j}(\overline{nw_j Q}) \tag{3.7}$$

and

$$\sum_j \iiint_{-\infty}^{\infty} Q(\mathbf{w}) F_j(\mathbf{r}, \mathbf{w}) \frac{\partial f}{\partial w_j} \, dw_x \, dw_y \, dw_z$$

$$= -\sum_j \iiint_{-\infty}^{\infty} f \frac{\partial}{\partial w_j} \{F_j(\mathbf{r}, \mathbf{w}) Q(\mathbf{w})\} \, dw_x \, dw_y \, dw_z$$

$$= -n \sum_j \overline{\frac{\partial}{\partial w_j} (F_j Q)}, \tag{3.8}$$

where we have integrated by parts in the first equality. We obtain the zeroth moment of (3.1) when $Q(\mathbf{w})$ is set equal to unity. The first two terms on the left-hand side reduce to $\partial n/\partial t$ and $\nabla \cdot (n\mathbf{v})$, respectively, and the last term is zero for forces for which $\partial F_j/\partial w_j = 0$. This condition is trivially true for all velocity-independent forces and is also true for (velocity-dependent) magnetic forces. Further, since collisions do not change the number density of particles, the integral of $(\partial f/\partial t)_{\text{coll}}$ over velocity space also vanishes. Thus, we obtain the equation of continuity of matter, viz.

$$\frac{\partial n}{\partial t} + \nabla \cdot (n\mathbf{v}) = 0 \tag{3.9}$$

or, with $\rho = nm$,

$$\frac{\partial \rho}{\partial t} + \nabla \cdot (\rho \mathbf{v}) = 0, \tag{3.9'}$$

where the particle density n is

$$n(\mathbf{r}, t) = \iiint_{-\infty}^{\infty} f(\mathbf{r}, \mathbf{w}, t) \, dw_x \, dw_y \, dw_z, \tag{3.10}$$

and the macroscopic velocity is

$$v(\mathbf{r}, t) = \bar{\mathbf{w}} = \frac{1}{n(\mathbf{r}, t)} \iiint_{-\infty}^{\infty} \mathbf{w} f(\mathbf{r}, \mathbf{w}, t) \, dw_x \, dw_y \, dw_z. \tag{3.11}$$

For later reference we note that an equation similar to the equation of continuity of matter (3.9') exists for the continuity of electric charge, ρ_e, viz.,

$$\frac{\partial \rho_e}{\partial t} + \nabla \cdot \mathbf{j} = 0. \tag{3.12}$$

Let us now set $Q(\mathbf{w}) = m\mathbf{w}$. This gives

$$\frac{\partial}{\partial t} (nm\mathbf{v}) + \nabla \cdot (nm\overline{\mathbf{w}\mathbf{w}}) - n\mathbf{F}$$

$$= \iiint_{-\infty}^{\infty} n\mathbf{w} \left(\frac{\partial f}{\partial t}\right)_{\text{coll}} dw_x \, dw_y \, dw_z. \tag{3.13}$$

The second term may be simplified using

$$\mathbf{w} = \mathbf{v} + \mathbf{u}, \tag{3.14}$$

where $\mathbf{u}$ is the random velocity component. This gives

$$\nabla \cdot (nm\overline{\mathbf{w}\mathbf{w}}) = \nabla \cdot (nm\overline{\mathbf{v}\mathbf{v}}) + \nabla \cdot (nm\overline{\mathbf{u}\mathbf{u}}), \tag{3.15}$$

with the cross-terms $\overline{\mathbf{u}\mathbf{v}}$ vanishing since $\bar{\mathbf{u}} = 0$ by definition. The second term in the right-hand side of (3.15) is the divergence of the stress tensor $\mathscr{P}$; the first term may be written

$$\nabla \cdot (nm\overline{\mathbf{v}\mathbf{v}}) = nm(\mathbf{v} \cdot \nabla)\mathbf{v} + \mathbf{v}(\nabla \cdot nm\mathbf{v}). \tag{3.16}$$

The left-hand side of (3.13) now becomes

$$nm\frac{\partial \mathbf{v}}{\partial t} + \mathbf{v}\frac{\partial(nm)}{\partial t} + nm(\mathbf{v} \cdot \nabla)\mathbf{v}$$
$$+ \mathbf{v}(\nabla \cdot (nm\mathbf{v})) + \nabla \cdot \mathscr{P} - n\mathbf{F}, \tag{3.17}$$

so that (3.13) becomes

$$nm\frac{d\mathbf{v}}{dt} = n\mathbf{F} - \nabla \cdot \mathscr{P} + \mathbf{P} \tag{3.18}$$

where $d/dt = (\partial/\partial t) + \mathbf{v} \cdot \nabla$ is the mobile operator and we have used the continuity equation (3.9′). $\mathscr{P}$ is the stress tensor which, in the simple case of an isotropic distribution of the random velocities of the particles, reduces to the scalar pressure p: $\nabla \cdot \mathscr{P} = \nabla p$, and $\mathbf{P}$ describes the momentum transfer by collisions, i.e., the right-hand side of (3.13). With an electric, $\mathbf{E}$, and magnetic, $\mathbf{B}_0$, field present in the plasma, the force $\mathbf{F}$ in (3.18) is $\mathbf{F} = g(\mathbf{E} + (\mathbf{v}/c) \times \mathbf{B}_0) + m\mathbf{g}$, where $\mathbf{g}$ is the acceleration due to gravity.

The general equation (3.18) may now be written for each species of the two-component (electron plus ion) plasma. Neglecting convective derivative terms, it takes the form

$$n_e m_e \frac{\partial \mathbf{v}_e}{\partial t} = -n_e e\left(\mathbf{E} + \frac{\mathbf{v}_e}{c} \times \mathbf{B}_0\right) - \nabla \cdot \mathscr{P}_e$$
$$+ n_e m_e \mathbf{g} + \mathbf{P}_{ei} \tag{3.19}$$

for the electrons, and

$$n_i m_i \frac{\partial \mathbf{v}_i}{\partial t} = n_i Ze\left(\mathbf{E} + \frac{\mathbf{v}_i}{c} \times \mathbf{B}_0\right) - \nabla \cdot \mathscr{P}_i$$
$$+ n_i m_i \mathbf{g} + \mathbf{P}_{ie} \tag{3.20}$$

for the ions. In order to obtain the bulk fluid behavior of the plasma, we add these equations together, giving

$$n_e m_e \frac{\partial \mathbf{v}_e}{\partial t} + n_i m_i \frac{\partial \mathbf{v}_i}{\partial t}$$

$$= n_e e \frac{(\mathbf{v}_i - \mathbf{v}_e) \times \mathbf{B}_0}{c} - \nabla p + (n_i m_i + n_e m_e)\mathbf{g}, \qquad (3.21)$$

where the interaction terms $\mathbf{P}_{ei}$ and $\mathbf{P}_{ie}$ are cancelled by Newton's third law. We have set $Zn_i = n_e$ by the neutrality conditions, and we have assumed a scalar pressure p. Note that the terms involving $\mathbf{E}$ cancel in a neutral plasma; the momentum gains by the electrons and ions in response to an applied electric field are necessarily equal and opposite. The terms involving $\mathbf{B}_0$ *do not* cancel; we are left with a term equal to $(\mathbf{j} \times \mathbf{B}_0)/c$, by (3.5).

The left-hand side of (3.21) is equal to $\rho(\partial \mathbf{v}/\partial t)$, by (3.3). Thus, (3.21) becomes

$$\rho \frac{\partial \mathbf{v}}{\partial t} = \frac{1}{c}\mathbf{j} \times \mathbf{B}_0 - \nabla p + \rho \mathbf{g}. \qquad (3.22)$$

It is the Lorentz force, $(1/c)\mathbf{j} \times \mathbf{B}_0$, which provides the all important coupling between the magnetic field and the material motions [see equation (3.5)].

Note that in fluid dynamics an equation somewhat different from (3.22) is used as the equation of motion. This is the Navier–Stokes equation where viscosity is included as a force term instead of gravity, i.e., $\rho \eta_v \nabla^2 \mathbf{v}$, where η_v is the coefficient of viscosity.

The energy transfer equation is the third conservation equation one generally needs to study the dynamic behavior of a plasma. To obtain this second moment we multiply (3.1) by w^2 and integrate over velocity space. However, in so doing we introduce third moments of the distribution function – the components of the heat flow tensor. If we can instead find an independent expression, from simpler physics, for the heating effect q, we will have a closed system of equations. In most plasmas, and certainly under solar flare conditions, the heating may be due to several sources. For example, if the sources are heat conduction, radiation, and Joule heating, the heat flow vector takes the forms

$$q = \nabla \cdot (K \nabla T), \quad -\nabla \cdot \mathbf{F}_{rad}, \quad \text{and} \quad \frac{j^2}{\sigma}, \qquad (3.23)$$

respectively, where K is the thermal conductivity, $\mathbf{F}_{rad}$ is the net radiative flux, and σ is the electrical conductivity. Note the different nature of these three expressions for q: while the last depends only on the local properties of the medium, the first two depend on the *global* structures of both the medium and the radiation field. Consequently, their effects are less straightforward to evaluate. Other examples of q will be considered in Chapters 4, 7, and 8. The energy equation is

$$\rho \frac{dU}{dt} = \frac{p}{\rho} \frac{d\rho}{dt} + q = -p\nabla \cdot \mathbf{v} + q, \qquad (3.24)$$

where U is the internal heat energy, i.e., the energy per unit mass.

In ordinary gas dynamics the heat flow is often so small that it can be ignored, and the energy conservation equation is simply an expression of the adiabatic variation of pressure, i.e., an equation for the conservation of entropy:

$$\frac{d}{dt}(p\rho^{-\gamma}) = 0. \qquad (3.25)$$

One will often see equation (3.25) used in solar atmospheric problems, notably at great depths where convection is a dominant mode of energy transport, but caution must be exercised whenever solar flare plasmas, which are associated with large changes in entropy, are involved.

3.2   The electrical conductivity and Ohm's law

The propagation of a disturbance in free space is completely and uniquely governed by Maxwell's equations, since only electromagnetic waves are possible under such conditions. However, in a plasma, i.e., in the presence of matter, a coupling exists between the radiation field and the particles. To describe such conditions we need an equation of motion for each kind of particle as well as Maxwell's equations. In such a description the concept of electrical conductivity will become important, and we shall therefore briefly discuss this concept first.

Let us return to the fundamental equations of motion (3.19) and (3.20). The electrical conductivity σ may be defined in terms of the momentum coupling between electrons and ions:

$$\mathbf{P}_{ei} = \frac{n_e e}{\sigma} \mathbf{j}; \qquad (3.26)$$

the proportionality between the momentum exchange rate $\mathbf{P}_{ei}$ and the drift current $\mathbf{j}$ is physically plausible. If we insert (3.26) into (3.19) and (3.20) with a scalar pressure, we obtain

$$n_e m_e \frac{\partial \mathbf{v}_e}{\partial t} = -n_e e \left(\mathbf{E} + \frac{\mathbf{v}_e}{c} \times \mathbf{B}_0 \right) - \nabla p_e$$

$$+ n_e m_e \mathbf{g} + \frac{n_e e}{\sigma} \mathbf{j} \qquad (3.27)$$

and

$$n_i m_i \frac{\partial \mathbf{v}_i}{\partial t} = n_i Z e \left(\mathbf{E} + \frac{\mathbf{v}_i}{c} \times \mathbf{B}_0 \right) - \nabla p_i$$

$$+ n_i m_i \mathbf{g} - \frac{n_e e}{\sigma} \mathbf{j}. \tag{3.28}$$

Having previously summed these equations to obtain the fluid equation of motion (3.22), we now seek information on the differential drift of electrons and ions within the moving fluid; i.e., on the currents present. Thus, taking $(m_e/e\rho) \times (3.28)$ minus $(m_i/eZ\rho) \times (3.27)$, and using the charge neutrality condition $Zn_i = n_e$, we obtain

$$\frac{m_i m_e}{Z \rho e^2} \left[Z n_i e \frac{\partial \mathbf{v}_i}{\partial t} - n_e e \frac{\partial \mathbf{v}_e}{\partial t} \right]$$

$$= \frac{n_e m_e + n_i m_i}{\rho} \mathbf{E} + \frac{(n_e m_e \mathbf{v}_i + n_i m_i \mathbf{v}_e) \times \mathbf{B}_0}{\rho c} + \frac{m_i}{eZ\rho} \nabla p_e$$

$$- \frac{m_e}{e\rho} \nabla p_i - \frac{n_e m_e + n_i m_i}{\rho \sigma} \mathbf{j}. \tag{3.29}$$

The term in parentheses in the second term on the right-hand side may be written

$$n_e m_e \mathbf{v}_e + n_i m_i \mathbf{v}_i + n_e m_e (\mathbf{v}_i - \mathbf{v}_e) - n_i m_i (\mathbf{v}_i - \mathbf{v}_e)$$

$$= \rho \mathbf{v} + \frac{(n_e m_e - n_i m_i)}{e n_e} \mathbf{j},$$

so that the following relation results:

$$\frac{m_i m_e}{Z \rho e^2} \frac{\partial \mathbf{j}}{\partial t} = \mathbf{E} + \frac{\mathbf{v} \times \mathbf{B}_0}{c} - \frac{\mathbf{j}}{\sigma}$$

$$+ \frac{1}{eZ\rho} \left[m_i \left(\nabla p_e - \frac{\mathbf{j} \times \mathbf{B}_0}{c} \right) - Z m_e \left(\nabla p_i - \frac{\mathbf{j} \times \mathbf{B}_0}{c} \right) \right], \tag{3.30}$$

where we have used the definition $\rho = n_e m_e + n_i m_i$, equation (3.5) for $\mathbf{j}$, and equation (3.3) for $\mathbf{v}$. This is the generalized form of Ohm's law. Note that the gravitational terms, which affect only bulk motion and not currents, play no role.

The term on the left-hand side of equation (3.30) is called the *inertial* term and corresponds to the transient change in a current when an electric field $\mathbf{E}$ is suddenly applied. This term may be important during the initial phase of, say, the injection of an electron beam into a plasma (cf. Chapters 4 and 7). However, after the current density $\mathbf{j}$ has grown as a result of this transient, collisional drag forces, proportional to $\mathbf{j}$, start to balance the electromotive force, so that a steady state is asymptotically obtained, with $\partial j/\partial t = 0$, and the left-hand side vanishes. Assuming force balance for the electrons, $\nabla p_e = (j \times B_0)/c$, and neglecting the term Zm_e compared to m_i, we obtain the steady state of Ohm's law:

$$j = \sigma\left(E \times \frac{v}{c} \times B_0\right), \tag{3.31}$$

or, in the absence of an external magnetic field, the familiar results from elementary physics,

$$j = \sigma E. \tag{3.32}$$

Equation (3.31) may also be written as

$$\eta j = E + \frac{1}{c}v \times B_0, \tag{3.33}$$

where η is the plasma resistivity. Equations (3.31) and (3.33) may be written

$$j = \sigma E_v, \tag{3.34}$$

and

$$\eta j = E_v, \tag{3.35}$$

where E_v is the electric field measured in a system moving with the velocity v of the plasma. It may contain an externally applied field E, as well as a field $(v/c) \times B_0$, induced by the motion.

The resistances to motion parallel and perpendicular to the magnetic field B_0 are not in general equal, being higher for currents perpendicular to B_0 [see equations (3.44) and (3.45) below]. In such cases, (3.34) and (3.35) are replaced more properly by

$$j = \sigma \cdot E_v, \tag{3.36}$$

and

$$\eta \cdot j = E_v, \tag{3.37}$$

where σ and η are tensors.

In the solar plasma, the resistivity η is often very small (σ very large). Under such conditions E_v is approximately zero (for j to remain finite) and we find

$$E \simeq -\frac{v}{c} \times B_0; \tag{3.38}$$

i.e., the electric field is approximately equal to the induced field.

In the simple scalar form of Ohm's law, the electrical conductivity is given by

$$\sigma = \frac{n_e e^2}{m_e v_c} \equiv \sigma_0, \tag{3.39}$$

where v_c is the collision frequency. This expression follows from the basic definition (3.26), setting $\mathbf{P}_{ei}$ (the rate of momentum loss per unit volume) $= n_e(m_e\mathbf{v})v_c$. Other simplified expressions exist, important among which is the conductivity σ_L of the hypothetical Lorentz gas, which is a fully ionized gas in which the ions are at rest and there are no electron–electron collisions. Under these conditions,

$$\sigma_L = \frac{2(2kT)^{3/2}}{\pi^{3/2}m_e^{1/2}Ze^2\ln\Lambda}$$

$$= 2.37 \times 10^8 \frac{T^{3/2}}{Z\ln\Lambda}\, s^{-1}, \tag{3.40}$$

where Λ is a slowly varying function of T and n_e:

$$\Lambda = \frac{\lambda_D}{\rho_0} = \frac{3}{2Ze^3}\left(\frac{k^3T^3}{\pi n_e}\right)^{1/2}. \tag{3.41}$$

Here λ_D is the *Debye length* or shielding distance

$$\lambda_D = \left(\frac{kT}{4\pi n_e e^2}\right)^{1/2}, \tag{3.42}$$

and indicates the distance to which a point charge can make its influence felt, or the distance within which strict electric neutrality in a plasma may not be obeyed; see Section 4.1.

The Lorentz gas requirement that no electron–electron collisions occur is unrealistic. When such collisions are included, we can write the expression for the conductivity in the form $\sigma_L' = \gamma_E\sigma_L$, where the correction factor, γ_E, which depends on the charge Z, varies monotonically from unity for an infinite charge ($Z \to \infty$) to 0.58 for $Z = 1$ (Spitzer, 1962).

Chapman and Cowling (1952) considered a slightly ionized gas in which close encounters determine the electron conductivity (in a fully ionized plasma, distant encounters between electrons and ions dominate) and derived the following expression for σ on the assumption that the particles may be considered rigid spheres.

$$\sigma = \text{const}\,\frac{\alpha e^2}{Q(m_e kT)^{1/2}}. \tag{3.43}$$

In expression (3.43) α is the degree of ionization and Q the collision cross-section for electron–atom interactions. In Section 4.4, we shall return to a more realistic treatment of collisions under the inverse-square Coulomb force and derive a more realistic formula for the electrical conductivity.

We now introduce a magnetic field $\mathbf{B}$ into a fully ionized plasma. The particles will gyrate in the field with the cyclotron, or Larmor, frequency ω_B

given by $\omega_B = eB/mc$. (For a derivation of this expression, see Section 4.2.)
Under these conditions the conductivity is a tensor and the current will have
three components. The conductivity for the current flowing along **B** is

$$\sigma_\parallel = \frac{n_e e^2}{m_e v_c} = \sigma_0.$$ (3.44)

We decompose the electric field into one component, $E_\parallel$, along **B**, and a
component $E_\perp$ at right-angles to **B**. Equation (3.44) shows that the magnetic
field has no influence on the parallel conductivity σ_0 [see (3.39)]. On the
other hand, the currents flowing perpendicular to the magnetic field, one
along $E_\perp$ and one perpendicular to both **B** and **E** (the *Hall current*), are
affected by the magnetic field. The current along $E_\perp$ has an electric
conductivity

$$\sigma_\perp = \frac{\sigma_0}{1 + \left(\dfrac{\omega_B}{v_c}\right)^2},$$ (3.45)

while the conductivity corresponding to the Hall current is

$$\sigma_H = \frac{\sigma_\perp}{\omega_B/v_c}.$$ (3.46)

We conclude this discussion of conductivity by noting that σ is intimately
related to the energy dissipated as heat by the currents. This Joule heating is
given by

$$W_J = \mathbf{j} \cdot \mathbf{E} = \frac{j^2}{\sigma} = \sigma E^2,$$ (3.47)

an expression which shows that, for a fixed current, when the conductivity is
very high, the heating is negligible. On the other hand, for a fixed electric
field, a high σ implies a high heating rate.

3.3   Maxwell's equations and electromagnetic waves

We have discussed in Section 3.2 the electric field vector, **E**, or
electric field intensity and the magnetic vector density, **B**, or magnetic
induction. These are the fundamental force vectors in free space. With the
introduction of the electric displacement, **D**, and the magnetic field
intensity, **H**, we can write Maxwell's equations in general form

$$\nabla \times \mathbf{H} = \frac{4\pi}{c}\mathbf{j} + \frac{1}{c}\frac{\partial \mathbf{D}}{\partial t},$$ (3.48)

$$\nabla \times \mathbf{E} = -\frac{1}{c}\frac{\partial \mathbf{B}}{\partial t},$$ (3.49)

$$\nabla \cdot \mathbf{D} = 4\pi\rho_e,$$ (3.50)

and

$$\nabla \cdot \mathbf{B} = 0, \tag{3.51}$$

where ρ_e is the electric charge density and $\mathbf{D}$ and $\mathbf{H}$ are derived vectors, associated with the state of matter, i.e., with the plasma present. Since $\mathbf{E}$ and $\mathbf{B}$ are our fundamental quantities, we would like to find expressions for these, which means that we need some additional information on $\mathbf{D}, \mathbf{H}$, and $\mathbf{j}$ to be able to eliminate them from the equations. In Section 3.2 we have seen [equation (3.32)] that under the simplified conditions discussed there, $\mathbf{j} = \sigma \mathbf{E}$. Further, one generally assumes a relationship

$$\mathbf{D} = \varepsilon \mathbf{E}, \tag{3.52}$$

where ε, the dielectric constant, in the general case, is a complicated tensor. Equation (3.52) is valid only if $\mathbf{D}$ is given by the local value of $\mathbf{E}$. It is mathematically convenient to combine σ and ε into a complex dielectric permittivity ε'

$$\varepsilon' = \varepsilon + i\left(\frac{4\pi\sigma}{\omega}\right), \tag{3.53}$$

where ω is the frequency of the assumed sinusoidal time dependence of all variables; for example, $E \sim \exp[i(\mathbf{k} \cdot \mathbf{r} - \omega t)]$. For a nonmagnetic medium the distinction between $\mathbf{B}$ and $\mathbf{H}$ vanishes. Using this information we find by combining equations (3.48) and (3.49)

$$\nabla^2 \mathbf{E} - \nabla\nabla \cdot \mathbf{E} - \frac{1}{c^2}\frac{\partial^2 \mathbf{D}}{\partial t^2} - \frac{4\pi}{c^2}\frac{\partial \mathbf{j}}{\partial t} = 0, \tag{3.54}$$

and inserting equations (3.32), (3.52), and (3.53) in (3.54) we arrive at the following basic wave equation for $\mathbf{E}$:

$$\nabla^2 \mathbf{E} - \nabla\nabla \cdot \mathbf{E} + \frac{\omega^2}{c^2}\varepsilon' \mathbf{E} = 0. \tag{3.55}$$

A similar equation can be derived for the magnetic induction vector $\mathbf{B}$. Equation (3.55) is very complicated, and different approximations lead to different expressions for ε and σ. For a wave propagating in the Z-direction, normal to a plane-parallel atmosphere, equation (3.55) reduces to

$$\frac{\partial^2 E_{x,y}}{\partial z^2} + \frac{\omega^2}{c^2}\varepsilon' E_{x,y} = 0, \tag{3.56}$$

a type of equation found in many areas of physics (e.g., acoustics; compare also to the Schrödinger equation). However, only specific, simplified cases can be treated analytically, since (3.56) has no solution in terms of known functions.

Reintroducing (3.53) into (3.55) we can write the basic equation for electromagnetic waves as

$$\nabla^2 \mathbf{E} - \nabla \nabla \cdot \mathbf{E} + \frac{\omega^2}{c^2} \varepsilon \mathbf{E} + i \frac{4\pi\omega}{c^2} \mathbf{j} = 0, \tag{3.57}$$

which, with $\rho_e = 0$ and the direction of propagation $\mathbf{k}$ in the Z-direction, becomes:

$$\frac{\partial^2 E_x}{\partial z^2} + \frac{\omega^2}{c^2} E_x + \frac{4\pi i \omega}{c^2} j_x = 0$$

and

$$\frac{\partial^2 E_y}{\partial z^2} + \frac{\omega^2}{c^2} E_y + \frac{4\pi i \omega}{c^2} j_y = 0. \tag{3.58}$$

In discussions of waves in a plasma, one encounters the concept of *Alfvén waves*, which are often treated as a special type of oscillation. This is certainly entirely appropriate, but as pointed out by Spitzer (1962), these waves can also be discussed in the framework of equation (3.57), which emphasizes the physics of these waves and shows that they also belong to the family of electromagnetic waves. In equation (3.58) we insert the following simplified forms of equations (3.22) and (3.30), viz.

$$\rho \frac{\partial v_z}{\partial t} = \frac{j_y B_0}{c}; \quad \text{i.e.,} \quad -i\omega\rho v_z = \frac{j_y B_0}{c},$$

and

$$E_y = \frac{v_z B_0}{c},$$

which results when we neglect time variations of $\mathbf{j}$, put $\mathbf{j} \times \mathbf{B} = 0$, and let $\sigma \to \infty$ [cf. equation (3.38)]. The equation governing the ensuing waves then takes the form

$$\frac{\partial^2 E_y}{\partial z^2} = \left(1 + \frac{4\pi\rho c^2}{B_0^2}\right) \frac{1}{c^2} \frac{\partial^2 E_y}{\partial t^2}. \tag{3.59}$$

Equation (3.59) is the equation for a wave through a plasma whose dielectric constant is

$$\varepsilon = 1 + \frac{4\pi\rho c^2}{B_0^2}. \tag{3.60}$$

The exceptionally high value of ε is therefore what characterizes an Alfvén wave when regarded as a normal electromagnetic wave. When $\varepsilon \gg 1$, equation (3.60) shows that the wave speed is

$$V_A = c\varepsilon^{-1/2} \approx \frac{B}{(4\pi\rho)^{1/2}}. \tag{3.61}$$

The complexity of general magnetohydrodynamics becomes evident when we look at the set of equations necessary to describe the physical situation in the plasma. To survey the problem we present the set of equations below, namely, the equations of continuity of mass and electric charge [(3.9) and (3.12)], the equations of motion [(3.19) and the similar one for ions], the energy equation [(3.24)], and the wave equation [(3.57)]:

$$\frac{\partial \rho}{\partial t} + \nabla \cdot (\rho \mathbf{v}) = 0, \tag{3.62}$$

$$\frac{\partial \rho_e}{\partial t} + \nabla \cdot \mathbf{j} = 0, \tag{3.63}$$

$$n_e m_e \frac{\partial \mathbf{v}_e}{\partial t} = -n_e e\left(\mathbf{E} + \frac{\mathbf{v}_e}{c} \times \mathbf{B}_0\right) - \nabla \cdot \mathscr{P}_e$$
$$+ n_e m_e \mathbf{g} + \mathbf{P}_{ei}, \tag{3.64}$$

$$n_i m_i \frac{\partial \mathbf{v}_i}{\partial t} = n_i Ze\left(\mathbf{E} + \frac{\mathbf{v}_i}{c} \times \mathbf{B}_0\right) - \nabla \cdot \mathscr{P}_i$$
$$+ n_i m_i \mathbf{g} + \mathbf{P}_{ie}, \tag{3.65}$$

$$\rho \frac{dU}{dt} = -p\nabla \cdot \mathbf{v} + \rho \mathbf{g}, \tag{3.66}$$

$$\nabla^2 \mathbf{E} - \nabla\nabla \cdot \mathbf{E} + \frac{\omega^2}{c^2} + i\frac{4\pi\omega}{c^2}\mathbf{j} = 0. \tag{3.67}$$

This set gives us 12 equations for the 12 quantities $\mathbf{E}, \mathbf{j}, \mathbf{v}, \mathbf{P}, \rho$, and T; a set that can theoretically be solved, but which in practice can only give useful results under drastically simplifying assumptions.

3.4 Waves in a plasma

A plasma, particularly a magnetized one such as the solar atmosphere, is capable of an extremely rich variety of wave-like motions. Frequently, high-level saturation of waves has important implications for the structure of the plasma. In addition, the response of a plasma to incoming waves, such as electromagnetic waves, is frequently of interest. In this section we shall discuss the properties of the more fundamental wave motions.

3.4.1 *Plasma oscillations*

If a quasi-neutral plasma is separated into positively and negatively charged parts, these parts will attract each other, providing a restoring force in the opposite direction to the original separation. The natural result of such a scenario is simple harmonic oscillations about the equilibrium (neutrality) configuration.

Consider the equations of continuity and momentum for the electrons in the plasma (the ions, being much more massive, are less likely to contribute to the oscillation), i.e., using equations (3.9) and (3.19) with $\mathbf{B}_0 = \mathbf{g} = \mathbf{P}_{ei} = \mathscr{P}_e = 0$,

$$\frac{\partial n_e}{\partial t} + \nabla \cdot (n_e \mathbf{v}_e) = 0,$$

$$m_e n_e \left(\frac{\partial \mathbf{v}_e}{\partial t} + (\mathbf{v}_e \cdot \nabla) \mathbf{v}_e \right) = -e n_e \mathbf{E},$$

plus Gauss's law in the form

$$\nabla \cdot \mathbf{E} = 4\pi \rho_e = 4\pi e (Z n_i - n_e).$$

We shall consider motions small enough that these equations can be linearized, yielding

$$\frac{\partial n_1}{\partial t} + n_0 \nabla \cdot \mathbf{v}_1 = 0, \tag{3.68}$$

$$m_e \frac{\partial \mathbf{v}_1}{\partial t} = -e \mathbf{E}_1, \tag{3.69}$$

and

$$\nabla \cdot \mathbf{E}_1 = -4\pi e n_1, \tag{3.70}$$

where the subscript '0' denotes equilibrium quantities ($\mathbf{v}_0 = 0$), and the subscript '1' denotes perturbed (small) quantities.

Any oscillation can be Fourier-decomposed into a sum of sinusoidal oscillations, and so it is sufficient to consider disturbances of the form

$$q_1 = q_1{}^* \exp[i(\mathbf{k} \cdot \mathbf{r} - \omega t)],$$

where $q_1{}^*$ is the amplitude of the disturbed quantity q_1. Dropping the asterisk hereafter, equations (3.68) through (3.70) give

$$-i\omega n_1 + n_0 i \mathbf{k} \cdot \mathbf{v}_1 = 0, \tag{3.71}$$

$$-i\omega m_e \mathbf{v}_1 = -e \mathbf{E}_1, \tag{3.72}$$

and

$$i\mathbf{k} \cdot \mathbf{E}_1 = -4\pi e n_1. \tag{3.73}$$

Multiplying equation (3.72) by $\mathbf{k}$ and eliminating $\mathbf{k} \cdot \mathbf{E}_1$ between this and (3.73), we find

$$-i\omega m_e \mathbf{k} \cdot \mathbf{v}_1 = -4\pi i e^2 n_1,$$

and elimination of $\mathbf{k} \cdot \mathbf{v}_1$ with (3.71) gives

$$\frac{\omega^2 m_e n_1}{n_0} = 4\pi e^2 n_1; \tag{3.74}$$

i.e.,

$$\omega = \left(\frac{4\pi n_0 e^2}{m_e}\right)^{1/2}. \tag{3.75}$$

This is the *electron plasma frequency*, denoted ω_{pe}. It depends only on the equilibrium density n_0 and has the numerical value $\omega_{pe}(s^{-1}) = 5.64 \times 10^4 [n_0(cm^{-3})]^{1/2}$. A cold plasma will oscillate with this frequency; however, since the group velocity $d\omega/dk = 0$, waves do not propagate by this mechanism.

If the plasma is warm, however, thermal motion of the electrons will carry information from one part of the plasma to another. The plasma oscillations now have a finite group velocity $v_g = d\omega/dk$ and represent a traveling wave. The dispersion relation for this wave is formed by adding a $-\nabla p_e$ term to the momentum equation (3.69) and continuing as before. With an isentropic energy equation $p_e \propto n_e^\gamma$, the linearized term $\nabla p_1 = \gamma k T_e \nabla n_1$, and equation (3.72) is replaced by

$$-i\omega m_e v_1 = -eE_1 - \frac{\gamma k T_e}{n_0} i n_1 k. \tag{3.76}$$

The final relation for ω is then

$$\omega^2 = \omega_{pe}^2 + \gamma k^2 v_e^2 = \omega_{pe}^2 (1 + \gamma k^2 \lambda_D^2), \tag{3.77}$$

where $v_e = (kT_e/m_e)^{1/2}$ is the electron thermal speed and λ_D is the Debye length (Section 4.1). The group velocity for such waves is finite;

$$v_g = \frac{d\omega}{dk} = \gamma v_e^2 \frac{k}{\omega} = \frac{\gamma v_e^2}{v_\phi}, \tag{3.78}$$

where v_ϕ is the phase velocity of the wave. Although the phase velocity can be arbitrarily large, the group velocity (at which information propagates) is always less than $\gamma^{1/2} v_e$.

Let us now consider wave motions associated with oscillations of ions. Since the ions are relatively massive, they do not have wave amplitudes large enough to actually collide with other ions, and they must use the intervening electron gas as a means of transmitting impulses. The linearized continuity equation for the ions is [cf. (3.71)]

$$\omega n_1 = n_0 k \cdot v_1. \tag{3.79}$$

The momentum equation is [cf. (3.76)]

$$-i\omega m_i v_1 = -eE_1 - \left(\frac{\gamma_e k T_e}{n_0} + \frac{\gamma_i k T_i}{n_0}\right) i n_1 k, \tag{3.80}$$

where γ_e and γ_i are the adiabatic indices for electrons and ions, respectively.

The replacement for Gauss's law (3.73) is more subtle. Because of the relatively large inertia of ions compared to electrons, any ion motion is immediately accompanied by a corresponding electron motion – a process known as *ambipolar diffusion*. Therefore, although oscillating electrons can cause a change in **E**, the relatively slow oscillations associated with ions do not. Equation (3.73) can therefore be replaced by

$$\mathbf{E}_1 = 0. \tag{3.81}$$

Taking $\mathbf{k} \cdot$ (3.80) and eliminating $\mathbf{k} \cdot \mathbf{v}_1$ using (3.79), we find

$$\omega^2 = k^2 \left(\frac{\gamma_e k T_e + \gamma_i k T_i}{m_i} \right), \tag{3.82}$$

for which the phase and group velocities are both equal to the sound speed

$$v_s = \left(\frac{\gamma_e k T_e + \gamma_i k T_i}{m_i} \right)^{1/2}. \tag{3.83}$$

Note that even if $T_i \to 0$, the thermal motions of the electrons (finite T_e) still carry the wave. The inertia term is still the ion mass, however. Because the thermal electrons responsible for carrying the wave travel a relatively large distance equal to $(m_i/m_e)^{1/2}$ wavelengths in one wave period, the wave can be considered isothermal, with $\gamma_e = 1$. Thus, in the limit $T_e \gg T_i$,

$$v_s = \left(\frac{k T_e}{m_i} \right)^{1/2} = c_s, \tag{3.84}$$

the *ion-acoustic speed*.

3.4.2   *Magnetoacoustic waves*

Up to this point we have ignored the possible (and, indeed, in the solar case, probable) presence of a magnetic field **B**, either internal to the plasma or associated with an external driver, such as in the plasma response to an electromagnetic wave. The simplest type of wave associated with a magnetic field is the Alfvén wave, obtained by ignoring all fields except **B**. We have seen in Section 3.3 how these waves may be considered a special case of electromagnetic waves [equation (3.57)], whose equation reduces to equation (3.59). We can also derive the characteristics of Alfvén waves by starting from the relevant equation of momentum

$$\rho \frac{\partial \mathbf{v}}{\partial t} = \frac{\mathbf{j} \times \mathbf{B}}{c} = \frac{1}{4\pi} (\nabla \times \mathbf{B}) \times \mathbf{B},$$

to which we add the frozen-in flux condition appropriate for a plasma of high conductivity [see equation (3.98)]

$$\frac{\partial \mathbf{B}}{\partial t} = \nabla \times (\mathbf{v} \times \mathbf{B})$$

and the Maxwell equation

$$\nabla \cdot \mathbf{B} = 0.$$

Linearizing these equations, we find

$$\rho_0 \frac{\partial \mathbf{v}_1}{\partial t} = \frac{1}{4\pi} (\nabla \times \mathbf{B}_1) \times \mathbf{B}_0,$$

$$\frac{\partial \mathbf{B}_1}{\partial t} = \nabla \times (\mathbf{v}_1 \times \mathbf{B}_0),$$

and

$$\nabla \cdot \mathbf{B}_1 = 0,$$

where we have taken an initially potential field: $\nabla \times \mathbf{B}_0 = 0$. Fourier analysis gives

$$-\mathrm{i}\omega \rho_0 \mathbf{v}_1 = \frac{\mathrm{i}}{4\pi} (\mathbf{k} \times \mathbf{B}_1) \times \mathbf{B}_0,$$

$$-\mathrm{i}\omega \mathbf{B}_1 = \mathrm{i}\mathbf{k} \times (\mathbf{v}_1 \times \mathbf{B}_0),$$

and

$$\mathbf{k} \cdot \mathbf{B}_1 = 0.$$

Expanding the vector triple products and noting that $\mathbf{k} \cdot \mathbf{B}_0 = 0$ ($\nabla \cdot \mathbf{B}_0 = 0$), we obtain the following set of equations defining the Alfvén waves:

$$\omega \mathbf{v}_1 = \frac{1}{4\pi\rho_0} (\mathbf{B}_0 \cdot \mathbf{B}_1) \mathbf{k}, \tag{3.85}$$

$$\omega \mathbf{B}_1 = (\mathbf{k} \cdot \mathbf{v}_1) \mathbf{B}_0, \tag{3.86}$$

and

$$\mathbf{k} \cdot \mathbf{B}_1 = 0. \tag{3.87}$$

Fig. 3.1. Torsional oscillations of a magnetic flux tube – a simple form of Alfvén wave.

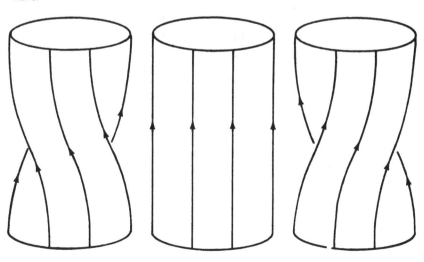

Multiplying equation (3.85) by **k** and equation (3.86) by $\mathbf{B}_0$ and eliminating the $(\mathbf{B}_0 \cdot \mathbf{B}_1)$ term, we arrive at the dispersion relation

$$\omega^2 = \frac{B_0^2}{4\pi\rho_0} k^2. \qquad (3.88)$$

This is another nondispersive wave, with phase and group velocities both equal to the Alfvén speed (3.61):

$$V_A = \frac{B_0}{(4\pi\rho_0)^{1/2}}.$$

Physically, the restoring force of an Alfvén wave is the tension in the magnetic field lines, with the fluid providing the inertia term. A simple form has the magnetic field $\mathbf{B}_0$ uniform along a cylinder, with **v** representing torsional oscillations of the field lines (Figure 3.1).

3.5   The magnetic diffusion equation

We conclude this chapter with a treatment of the behavior of the magnetic field in a plasma under certain simplifying conditions, of great importance for flare physics.

In astrophysical plasmas, as opposed to those in the terrestrial laboratory, the magnetic field plays a dominant role. Currents are produced from magnetic fields through Ampère's law; Ohm's law, in the presence of a finite resistivity, is responsible for the establishment of electric fields. Contrast this with the usual situation in the laboratory, where the electric field (battery) is the *ab initio* driver, controlling the current through Ohm's law, with the magnetic field following, through Ampère's law, from this current. It is important to realize this distinction between astrophysical and laboratory electromagnetic fields – as we shall see in this section, it is a consequence of the much larger size scales in the astrophysical environment. The magnetic diffusion equation is the foundation for the above statements, and we now derive this fundamental equation.

We start with Maxwell's equations for an unmagnetized, electrically neutral plasma, i.e., equations (3.48) and (3.49) together with Ohm's law, equation (3.33), viz.

$$\nabla \times \mathbf{B} = \frac{4\pi}{c}\mathbf{j},$$

$$\nabla \times \mathbf{E} = -\frac{1}{c}\frac{\partial \mathbf{B}}{\partial t},$$

and

$$n\mathbf{j} = \mathbf{E}_v = \mathbf{E} + \frac{\mathbf{v} \times \mathbf{B}_0}{c}.$$

In Ampère's law (3.48) we have neglected the displacement current term $(1/c)\,\partial \mathbf{E}/\partial t$. This follows from simple dimensional arguments under the valid assumption that fluid velocities are much less than c. The term $c^{-1}\,\partial \mathbf{E}/\partial t \approx (v/c)(\mathbf{E}/l)$, where l is a characteristic length scale associated with fluid motions. By comparison $\nabla \times \mathbf{B} \approx \mathbf{B}/l \approx (c/v)(\mathbf{E}/l)$, by (3.49); thus, the displacement current term is smaller by (v^2/c^2) than the left-hand side of (3.48) and can be safely neglected. The same argument allows us to neglect the relativistic Lorentz factor in the transformation (3.33).

Substitution for $\mathbf{E}$ in Faraday's law (3.49) from Ohm's law (3.33) gives

$$\frac{\partial \mathbf{B}}{\partial t} = \nabla \times (\mathbf{v} \times \mathbf{B}) - c\nabla \times (\eta \mathbf{j}), \tag{3.89}$$

and eliminating $\mathbf{j}$ between this and Ampère's law (3.48), we find

$$\frac{\partial \mathbf{B}}{\partial t} = \nabla \times (\mathbf{v} \times \mathbf{B}) - \frac{\eta c^2}{4\pi} \nabla \times (\nabla \times \mathbf{B}) - \frac{c^2}{4\pi} \nabla \eta \times (\nabla \times \mathbf{B}). \tag{3.90}$$

To simplify matters somewhat, we shall assume approximately uniform resistivity over the characteristic size scale l mentioned above; then the last term in (3.90) can be neglected. Use of the vector identity

$$\nabla \times (\nabla \times \mathbf{B}) = \nabla(\nabla \cdot \mathbf{B}) - \nabla^2 \mathbf{B},$$

and Gauss' law for magnetic fields (3.51),

$$\nabla \cdot \mathbf{B} = 0,$$

renders (3.90) in the form

$$\frac{\partial \mathbf{B}}{\partial t} = \nabla \times (\mathbf{v} \times \mathbf{B}) + \frac{\eta c^2}{4\pi} \nabla^2 \mathbf{B}, \tag{3.91}$$

which is the magnetic diffusion equation. The last term clearly has the form of a diffusive term, with the rate of diffusion characterized by the plasma resistivity η. We can more formally understand the meaning of (3.91) by considering the magnetic flux enclosed by a contour C

$$\Phi = \iint_S \mathbf{B} \cdot d\mathbf{S}, \tag{3.92}$$

where $\mathbf{S}$ is a surface bounded by the contour. Let the contour C now represent a closed contour *moving with the fluid*. After some time dt the contour will be displaced to C' (see Figure 3.2). Any change in the magnetic flux through the contour will be due to two causes: the temporal change in $\mathbf{B}$ during the time the contour moves from C to C', and the flux lost through the 'walls' defined by the fluid streamlines. The area 'along the walls' swept out by an element of C in moving to C' is

$$dS_W = dl \times v \, dt; \tag{3.93}$$

thus, we find that the change in magnetic flux is

$$d\Phi = \iint_S \frac{\partial \mathbf{B}}{\partial t} \cdot dS \, dt - \oint_C \mathbf{B} \cdot (dl \times v \, dt). \tag{3.94}$$

Using the identity

$$\mathbf{B} \cdot (dl \times v) = dl \cdot (v \times \mathbf{B}), \tag{3.95}$$

Fig. 3.2. A surface S bounded by a contour C and moving with the plasma. After a small time interval dt, the surface has deformed to S', bounded by C'. The difference in the magnetic flux through S' from that through S arises from two causes – a change in the strength of $\mathbf{B}$ and the escape of flux through the walls defined by the fluid streamlines. The area vector dS_w for the wall element shown is equal to $dl \times v \, dt$.

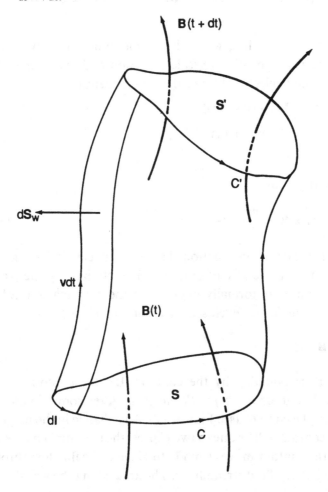

we may express (3.94) in the form

$$\frac{d\Phi}{dt} = \iint_S \frac{\partial \mathbf{B}}{\partial t} \cdot d\mathbf{S} - \oint_C (\mathbf{v} \times \mathbf{B}) \cdot d\mathbf{l}, \tag{3.96}$$

and, using Stokes' theorem to express the last term as a surface integral over S, we arrive at

$$\frac{d\Phi}{dt} = \iint_S \left[\frac{\partial \mathbf{B}}{\partial t} - \nabla \times (\mathbf{v} \times \mathbf{B})\right] \cdot d\mathbf{S}. \tag{3.97}$$

Thus, by the magnetic diffusion equation (3.91),

$$\frac{d\Phi}{dt} = \frac{\eta c^2}{4\pi} \iint_S \nabla^2 \mathbf{B} \cdot d\mathbf{S}. \tag{3.98}$$

We thus see that in a perfectly conducting plasma ($\eta = 0$), Φ is constant in time; i.e., *the flux through any surface bounded by a contour moving with the plasma does not change.* We say that the field lines are 'frozen-in' to the plasma, such that the two are always constrained to move in unison. This fundamental result, first derived by Alfvén (1950) [for a somewhat more rigorous proof, see Parker (1979), pp. 34–35] can be understood as follows: in a zero resistivity plasma, any finite emf $\mathbf{E}$ will, by Ohm's law, drive an infinite current. Since this is physically impossible, $\mathbf{E}$ is constrained to be zero. However, Faraday's law is expressed physically by saying that an emf is generated whenever a circuit element 'cuts through lines of magnetic flux'; since such emfs are forbidden by our earlier statement, the plasma must move with the lines of flux and vice versa.

To see when the highly conducting assumption is valid, we return to the magnetic diffusion equation (3.91) and define the *magnetic Reynolds number*, R_m, by

$$R_m = \frac{|\nabla \times (\mathbf{v} \times \mathbf{B})|}{\left|\dfrac{\eta c^2}{4\pi} \nabla^2 \mathbf{B}\right|}, \tag{3.99}$$

i.e., the ratio of the convective to diffusive terms in the right-hand side of (3.91). The size of R_m determines to what extent the field is frozen into the plasma: the higher R_m, the better the approximation.

Let the scale length for variation of B be l. Then, to order of magnitude,

$$R_m = \frac{4\pi v l}{\eta c^2}. \tag{3.100}$$

Let us evaluate the magnetic Reynolds number for two cases – a copper wire in the laboratory and the pre-flare solar corona. In the former case, $\eta_{copper} \approx 2 \times 10^{-18}$ esu, $l \approx 1\,\mathrm{cm}$, and the current drift velocity $v \approx 0.1\,\mathrm{cm\ s^{-1}}$ (for a

1 A current over a wire of cross-section $1 \, cm^2$). Thus R_m in this case has a value of about 10^{-5}, and we conclude that the frozen-in approximation is a very poor one. In the ionized plasma of the solar corona, $\eta = 1.5 \times 10^{-7} T^{-3/2}$ [see equations (3.40) and (4.83)], which for $T \approx 3 \times 10^6 \, K$ gives $\eta \approx 3 \times 10^{-17} \, esu$, comparable to the copper wire. However, the relevant v is now the Alfvén speed $V_A \approx 2 \times 10^9 \, cm \, s^{-1}$ (for a $10^{10} \, cm^{-3}$ plasma and a $1000 \, G$ field [see equation (3.61)], and so

$$R_m \approx 10^6 l, \tag{3.101}$$

which is $\gg 1$ for any reasonable l. Thus, the frozen-in approximation is extremely good for these conditions. We see that this is largely a consequence of the large lengths and velocities as compared to laboratory conditions.

Although useful in enabling us to use radiating plasma as a tracer of magnetic field lines in the Sun's atmosphere, the high value of R_m makes it extremely difficult to release the energy stored in stressed magnetic fields. Such an energy-release process proceeds over a diffusion timescale τ_D, given by considering the second term on the right-hand side of (3.91), so that

$$\tau_D \approx \frac{B}{(\partial B/\partial t)} \approx \frac{4\pi l^2}{\eta c^2}. \tag{3.102}$$

For the solar resistivity mentioned above and with $l \approx 10^9 \, cm$, the characteristic size of a flaring loop, τ_D, evaluates to $\approx 5 \times 10^{14} \, s$, i.e., about 10^7 years! Contrasting this with τ_D for the copper wire situation for which $\tau_D \approx 10^{-6} \, s$, we see the justification for the remarks made at the beginning of this section, viz. in the laboratory a magnetic field must be sustained through externally driven currents or it will leak away on very short timescales; whereas, in the astrophysical environment, magnetic fields persist for a long period and require no external means to sustain them.

3.6   Force-free fields

Energy can be built up in a magnetic field by stressing the field configuration so that it is no longer potential. In the solar atmosphere this buildup presumably occurs on a timescale much longer than the energy-release time, as subphotospheric motions act on potential field configurations to produce a stressed configuration. Since the associated motions have an associated velocity which is much smaller than the characteristic velocity (in this case, the Alfvén velocity), we may approximate the magnetic field at each instant by a static configuration.

Consider, therefore, the equation of motion (3.22):

$$\rho \frac{\partial \mathbf{v}}{\partial t} = \frac{1}{c} \mathbf{j} \times \mathbf{B} - \nabla p + \rho \mathbf{g}. \tag{3.22}$$

In a steady state, $\partial \mathbf{v} / \partial t = 0$ and

$$\frac{\mathbf{j} \times \mathbf{B}}{c} = \nabla p - \rho \mathbf{g}, \tag{3.103}$$

or, using Ampère's law (3.48),

$$\frac{1}{4\pi} (\nabla \times \mathbf{B}) \times \mathbf{B} = \nabla p - \rho \mathbf{g}. \tag{3.104}$$

Using the vector identity

$$\nabla(\mathbf{A} \cdot \mathbf{B}) = \mathbf{A} \times (\nabla \times \mathbf{B}) + \mathbf{B} \times (\nabla \times \mathbf{A}) + (\mathbf{A} \cdot \nabla)\mathbf{B}$$
$$+ (\mathbf{B} \cdot \nabla)\mathbf{A},$$

with $\mathbf{A} = \mathbf{B}$, we find that (3.104) may be written

$$\frac{1}{4\pi} (\mathbf{B} \cdot \nabla)\mathbf{B} = \nabla \left(\frac{B^2}{8\pi} + p \right) + \rho \nabla \Phi_\mathrm{g}, \tag{3.105}$$

where Φ_g is the gravitational potential. If we consider the magnitudes of $B^2/8\pi$, p, and Φ_g for the solar atmosphere, we find that both p and Φ_g are much smaller than $B^2/8\pi$, so that, to a very good approximation, we may neglect the gas pressure and gravitational terms, and so replace (3.104) by the *force-free field equation*

$$(\nabla \times \mathbf{B}) \times \mathbf{B} = 0. \tag{3.106}$$

This condition, which physically signifies a zero Lorentz force $\mathbf{j} \times \mathbf{B}$ (there being no other forces of comparable magnitude to balance a finite one), can be satisfied in three ways: $\mathbf{B} = 0$ (trivial), $\nabla \times \mathbf{B} = 0$ (i.e., $\mathbf{B} = \nabla \phi$, the condition for a potential field), or

$$\nabla \times \mathbf{B} = \alpha \mathbf{B}, \tag{3.107}$$

where the scalar $\alpha = \alpha(\mathbf{r})$ in general. This last case is the most interesting physically. It represents a stressed (nonpotential) field with its current $\mathbf{j}$ parallel to $\mathbf{B}$ everywhere, so that the flow of charged particles does not cross any field lines, thereby avoiding large $\mathbf{j} \times \mathbf{B}$ forces.

Taking the divergence of (3.107) and noting that $\nabla \cdot \mathbf{B} = 0$ and $\nabla \cdot (\nabla \times \mathbf{B}) = 0$ identically, we find that

$$\mathbf{B} \cdot \nabla \alpha = 0. \tag{3.108}$$

Equation (3.108) says that the gradient of α is always perpendicular to $\mathbf{B}$, so that α is constant along any field line. In the special case where α is a constant everywhere ($\nabla \alpha = 0$), we find that (3.107) is a linear equation, whose solution is therefore considerably simplified. We shall return to the force-free field equation in our study of pre-flare conditions in Chapter 6.

3.7   Magnetic buoyancy

Let us return to equation (3.105) and write it in the form

$$\frac{1}{4\pi}(\mathbf{B}\cdot\nabla)\mathbf{B} = \nabla\left(\frac{B^2}{8\pi}+p\right)-\rho\mathbf{g}. \tag{3.109}$$

Consider a flux tube of uniform field strength $\mathbf{B}=B_0\hat{\mathbf{x}}$, placed in a field-free atmosphere, plane stratified in the Z-direction. The left-hand side of equation (3.109) vanishes in this situation. If the tube is in pressure balance with its surroundings, $(p+B^2/8\pi)$ has the same value inside and outside the tube; consequently, the gas pressure p_i inside the tube is less than that outside (p_o) by an amount $B_0{}^2/8\pi$. If we further insist that the tube is in thermal equilibrium with its surroundings, then

$$2n_ikT = p_i = p_o - \frac{B_0{}^2}{8\pi} = 2n_okT - \frac{B_0{}^2}{8\pi}, \tag{3.110}$$

so that

$$n_i = n_o - \frac{B_0{}^2}{16\pi kT}. \tag{3.111}$$

The tube is therefore lighter than its surroundings and experiences an upthrust force of

$$F = (n_o - n_i)m_H gV = \frac{B_0{}^2 m_H gV}{16\pi kT}, \tag{3.112}$$

where m_H is the hydrogen mass and V the volume of the tube. The quantity $2kT/m_H g$ is the scale height H of the ambient atmosphere; thus,

$$F = \frac{B_0{}^2 V}{8\pi H}. \tag{3.113}$$

After the flux tube has risen a height H, it has acquired a kinetic energy

$$K = F\cdot H = \frac{B_0{}^2 V}{8\pi} = \tfrac{1}{2}\rho Vu^2, \tag{3.114}$$

where ρ is the tube's density and u is the final velocity. Solving for u, we find

$$u = \left(\frac{B_0{}^2}{4\pi\rho}\right)^{1/2} = V_A, \tag{3.115}$$

i.e., the Alfvén speed for the field strength B_0.

We therefore see that flux tubes have a tendency to rise at an appreciable speed. For a more rigorous treatment, including variations of temperature and pressure with height, we refer the reader to Parker (1979). This rising is eventually hindered by the tension acting on the ends of the flux tube in the solar interior, and we expect to see the field emerge as a loop-like structure. Indeed, as we shall see in Chapter 6, the magnetic loop is the fundamental building block of solar activity, in particular, flares, and we shall further see in Chapters 6 and 7 how stressing and interacting of these loops can lead to flare energy storage and release.

4

Flare plasma physics

No snowflake in an avalanche ever feels responsible

More Unkempt Thoughts, S. J. Lee (1909–66)

We saw in Chapter 3 that when a macroscopic, 'fluid' point of view could be applied to the treatment of solar atmospheric phenomena, the appropriate tool is provided by magnetohydrodynamics. However, solar flare conditions often exist where a microscopic point of view is necessary, in which we need to consider individual particles. We refer to such cases as plasma physics; although, to some extent, magnetohydrodynamics is also contained therein.

The subject of plasma physics is a complex one, and it is possible to devote volumes of material to its study. In this chapter, we will try to highlight some of the more pertinent areas which relate to the solar flare and which will be useful in later chapters. Excellent texts exist for the reader who wishes to pursue the subject further [e.g., Schmidt (1969); Krall and Trivelpiece (1973); Chen (1974)].

4.1   What is a plasma?

If one heats a gas to sufficiently high temperatures that the atoms completely ionize, the properties of the resulting gas will be principally controlled by the electromagnetic forces among the constituent ions and electrons. With the proviso that certain other properties, to be discussed shortly, are also satisfied, such a gas is called a *plasma*. This stripping of electrons from ions results in such a qualitatively different behavior that a plasma is sometimes called the 'fourth state of matter' (after solid, liquid, and gas).

One of the most notable properties of the Coulomb force between charged particles is its long range. Consider a volume of particles all of charge $+e$, say, and place a test particle of charge $+e$ at one end of the volume. For definiteness, let us consider the volume to be an elementary conical segment, with the test charge at the apex, as shown in Figure 4.1. The force due to the field particles within a shell bounded by distances r and $r + dr$ from the test charge is

$$dF = \frac{e^2}{r^2} (nr^2 \, dr \, d\Omega), \tag{4.1}$$

where the first factor is the Coulomb law between two particles, and the second factor is the number of field particles involved, $d\Omega$ being the solid angle of the cone in Figure 4.1. Expression (4.1) reduces to

$$dF = ne^2 \, dr \, d\Omega, \tag{4.2}$$

so that the force exerted by a shell is independent of the radius of the shell. Integration of (4.2) from $r = 0$ to ∞ would result in an infinite force on the test particle. (A similar argument based on the inverse square law of light propagation implies that the sky brightness in an infinite universe would be infinitely bright and is known as Olbers' paradox.)

The situation discussed above is artificial in that the field particles were all considered to have the same charge $+e$. If negatively charged field particles also were present, the net force on an elementary volume would be correspondingly reduced. In fact, if the net charge density in a given spherical shell were zero, dF for that shell also would be zero. More importantly however, the effect of placing the test charge $+e$ into such a neutral medium is to disturb the electrical neutrality of the medium. Negatively charged field particles will be attracted to the test charge, and positively charged field particles repelled from it, until the charge separation in the plasma creates an electric field equal and opposite to that of the test

Fig. 4.1. Conical element of plasma filled with charged particles of density n. The volume of the radial segment shown is $r^2 \, dr \, d\Omega$.

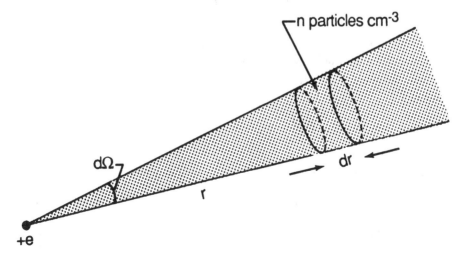

particle at the origin. As a result of this redistribution of charge, the distant field particles do not feel the full charge of the test particle, and we say that its effects are 'shielded'. This shielding property is characteristic of plasmas, and effectively confines the volume over which equation (4.1) is valid to within a 'shielding length' of the test charge. We now proceed to calculate this shielding length.

Suppose the field particles have charges $+e$ and $-e$, masses m_+ and m_-, temperatures T_+ and T_-, and unperturbed densities $n_{+,0} = n_{-,0} = n$. Then the undisturbed distribution function of speeds $f(v)$ for the field particles will be

$$f_{+,0}(v_+) = A_+ \exp\left[\frac{-m_+ v_+^2}{2kT_+}\right],$$ (4.3)

$$f_{-,0}(v_-) = A_- \exp\left[\frac{-m_- v_-^2}{2kT_-}\right],$$ (4.4)

where A_+ and A_- are normalization constants proportional to n and dependent on T_+ and T_-. If we now introduce our test charge $+e$ at the origin, each field particle will feel an additional electrostatic potential ϕ, which satisfies Poisson's equation

$$\nabla^2 \phi = -4\pi e(n_+ - n_-),$$ (4.5)

where n_+ and n_- are the perturbed densities of positively and negatively charged field particles respectively. The presence of this electrostatic potential gives each field particle an additional potential energy $\pm e\phi$, which modifies the distribution functions f_+ and f_- to the forms

$$f_+(v_+) = A_+ \exp\left[\frac{-(\frac{1}{2}m_+ v_+^2 + e\phi)}{kT_+}\right]$$ (4.6)

$$f_-(v_-) = A_- \exp\left[\frac{-(\frac{1}{2}m_- v_-^2 - e\phi)}{kT_-}\right].$$ (4.7)

As the distance r from the test charge increases to infinity, ϕ will go to zero as a result of complete shielding, and the distribution functions $f_+(v_+)$ and $f_-(v_-)$ will be given by their undisturbed values (4.3) and (4.4). Thus, we may write

$$f_+(v_+) = f_{+,0}(v_+)\exp\left(-\frac{e\phi}{kT_+}\right),$$ (4.8)

$$f_-(v_-) = f_{-,0}(v_-)\exp\left(\frac{e\phi}{kT_-}\right).$$ (4.9)

Integrating these distributions over v_+ and v_-, respectively, we find that

$$n_+ = n\exp\left(-\frac{e\phi}{kT_+}\right); \quad n_- = n\exp\left(\frac{e\phi}{kT_-}\right),$$ (4.10)

so that in the vicinity of a potential ϕ the number of ions is reduced, and the number of electrons increased, as expected. Poisson's equation (4.5) now becomes

$$\nabla^2\phi = -4\pi en\left[\exp\left(-\frac{e\phi}{kT_+}\right) - \exp\left(\frac{e\phi}{kT_-}\right)\right]. \qquad (4.11)$$

This equation is nonlinear and difficult to solve in general. However, if the energy due to the perturbing potential $e\phi$ is much less than the thermal energies kT_+ and kT_-, we may expand the exponentials in equations (4.10) and (4.11) and obtain, to first order, the linear equation

$$\nabla^2\phi = \frac{4\pi ne^2}{k}\left(\frac{1}{T_+} + \frac{1}{T_-}\right)\phi. \qquad (4.12)$$

In the case of spherical symmetry, the solution of (4.12) is

$$\phi = \frac{e^2}{r}e^{-r/\lambda_D}, \qquad (4.13)$$

where λ_D is the *Debye length*:

$$\lambda_D{}^{-2} = \frac{4\pi ne^2}{k}\left(\frac{1}{T_+} + \frac{1}{T_-}\right) = 0.021n\left(\frac{1}{T_+} + \frac{1}{T_-}\right). \qquad (4.14)$$

The solution (4.13) has been chosen so as to satisfy the condition $\phi \to 0$ as $r \to \infty$, and $\phi \to e^2/r$ (the standard Coulomb potential) as $r \to 0$.

Expression (4.13) shows that substantial deviations from neutrality in the field particles exist only for $r < \lambda_D$; for $r \gg \lambda_D$, $\phi \approx 0$ and the field particles are undisturbed (i.e., the gas remains 'quasi-neutral'). The required 'shielding length' is λ_D. In order for electromagnetic forces to control the structure of the gas, the volume defined by $r \leqslant \lambda_D$ must therefore contain a large number of particles N_D; i.e.,

$$N_D = \tfrac{4}{3}\pi n\lambda_D{}^3 \gg 1, \qquad (4.15)$$

and the characteristic dimension of the plasma L must satisfy the condition

$$L \gg \lambda_D. \qquad (4.16)$$

In such situations the ionized gas is controlled, at least locally, by long-range Coulomb forces and can justly be called a plasma. In order to study the motion of our test particle we must consider interactions not only with their nearest neighbors (binary collisions), but also with all the N_D particles within the 'Debye sphere' of radius $r = \lambda_D$ (the so-called 'collective interactions'). Thus, we are led to the formal definition of a plasma:

> *A plasma is a quasi-neutral gas of ionized particles which exhibits collective behavior.*

For the pre-flare solar corona, $T_+ = T_- \approx 2 \times 10^6$ K and $n \approx 10^9$ cm^{-3}.

Thus, from equation (4.14), $\lambda_D \approx 0.2$ cm and $N_D = \frac{4}{3}\pi n \lambda_D{}^3 \approx 4 \times 10^7$. Clearly, conditions (4.15) and (4.16) are both satisfied and we may treat the solar atmosphere as a true plasma. This means that its evolution on a microscopic scale is governed by a myriad of forces, far too numerous to consider individually. We are therefore forced to treat the plasma statistically and discuss its properties using the particle velocity distribution function $f(\mathbf{v})$. The exception to this is where the electrostatic field $e/\lambda_D{}^2$ at the limit of the Debye sphere is much smaller than some applied external field; in such cases, we may treat the electromagnetic fields are prescribed, rather than determined through internal self-consistency, and consider the motion of individual particles under the action of these prescribed fields. Let us now examine such cases.

4.2   Single-particle motions

The force on a particle of charge q and velocity $\mathbf{v}$ in prescribed $\mathbf{E}$ and $\mathbf{B}$ fields is given by the Lorentz equation

$$\mathbf{F} = q\left(\mathbf{E} + \frac{\mathbf{v} \times \mathbf{B}}{c}\right). \tag{4.17}$$

The case $\mathbf{B} = 0$ gives a uniform acceleration $\mathbf{a} = q\mathbf{E}/m$. The case $\mathbf{E} = 0$ results in a helical motion about the field lines: if we separate $\mathbf{v}$ into components $v_\parallel$ and $v_\perp$ along and perpendicular to $\mathbf{B}$, we derive the expressions

$$ma_\parallel = 0,$$
$$ma_\perp = \frac{qvB}{c}. \tag{4.18}$$

Because a is perpendicular to both $\mathbf{v}$ and $\mathbf{B}$, it corresponds to an inward (centripetal) acceleration; thus,

$$\frac{mv^2}{r_B} = \frac{qvB}{c}, \tag{4.19}$$

where r_B is the radius of the resulting orbit (the gyroradius). The solution of (4.19) for r_B gives

$$r_B = \frac{mvc}{qB}, \tag{4.20}$$

a formula valid in both nonrelativistic and relativistic regimes. The orbital frequency

$$\omega_B = \frac{v}{r_B} = \frac{qB}{mc} \tag{4.21}$$

is called the *gyrofrequency*; it is proportional to B and to the charge-to-mass ratio of the orbiting particle.

The combination of parallel and perpendicular motions gives rise to a helical motion of the particle; the sense of the helix depends on the charge of the particle and on the sense of $v_\parallel$ relative to **B**.

The addition of a weak electric field **E** to the spiraling particle causes it to be accelerated at some parts of its orbit and decelerated at others, as a result of the component of **E** perpendicular to **B**. It is also accelerated uniformly by the component parallel to **B**. Since $r \propto v$ [equation (4.20)], the resulting orbit deforms into a drifting orbital motion, as shown in Figure 4.2. Note that the direction of drift is the same for both positively and negatively charged particles.

To obtain a quantitative measure of the drift of the charged particle orbit, we recall from relativity theory that electric and magnetic fields are frame-dependent quantities, whose values are different for observers in relative motion. However, the quantity $E^2 - B^2$ is an invariant quantity in special relativity. For weak electric fields ($|E| < |B|$), $E^2 - B^2$ is negative, and it is possible to find a reference frame in which **E** vanishes, leaving only **B** and the simple motion of equation (4.18). To find this frame, we use the transformation equations for **E** and **B** in the form

$$\mathbf{E}_\parallel' = \mathbf{E}_\parallel,$$
$$\mathbf{E}_\perp' = \gamma\left(\mathbf{E}_\perp + \frac{\mathbf{V} \times \mathbf{B}}{c}\right), \tag{4.22}$$

Fig. 4.2. **E** × **B** drift. A magnetic field **B** points into the paper, causing positively charged particles to orbit counterclockwise and negatively charged particles to orbit clockwise [equation (4.17)]. An electric field **E** is then imposed, as shown by the arrow, accelerating the positive charges downward and the negative charges upward. As a particle gains or loses velocity, the radius of its orbit increases or decreases [equation (4.20)], producing the paths shown. Note that both charges drift in the same direction, a result of the covariance of the electromagnetic field [equation (4.25)].

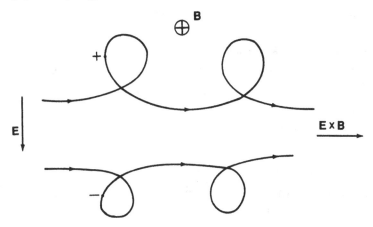

where $E_\parallel$ and $E_\perp$ are the components of E parallel and perpendicular to B, respectively, V is the velocity of the primed frame relative to the unprimed frame, and $\gamma \equiv (1 - V^2/c^2)^{-1/2}$ is the relativistic Lorentz factor. The $E_\parallel$ component is unchanged by the transformation and causes a uniform acceleration along the magnetic field lines. We can make E_1' vanish by requiring that

$$\mathbf{E} + \frac{\mathbf{V} \times \mathbf{B}}{c} = 0. \tag{4.23}$$

The vector product of this equation with B yields

$$\mathbf{E} \times \mathbf{B} + \frac{(\mathbf{V} \times \mathbf{B}) \times \mathbf{B}}{c} = 0,$$

and, expanding the vector triple product, we find

$$\mathbf{E} \times \mathbf{B} + \frac{1}{c}[(\mathbf{B} \cdot \mathbf{V})\mathbf{B} - B^2\mathbf{V}] = 0. \tag{4.24}$$

Without loss of generality, V can be taken to be perpendicular to B [cf. equation (4.22)], so that the first term in the square bracket vanishes, and

$$\mathbf{V} = c \frac{\mathbf{E} \times \mathbf{B}}{B^2}. \tag{4.25}$$

This is the drift velocity of the guiding center of the charged particle orbit. Note that it is indeed independent of the nature (charge, mass) of the charged particle, since it follows simply from the covariance of the electromagnetic field and not from the properties of the test particles placed in that field. In the guiding center frame $E_\perp$ vanishes and

$$\mathbf{B}' = \gamma\left(\mathbf{B} - \frac{\mathbf{V} \times \mathbf{E}}{c}\right) = \gamma\left(1 - \frac{E^2}{B^2}\right)\mathbf{B}. \tag{4.26}$$

The motion in this frame is a simple orbital motion, with altered gyrofrequency ω_B, resulting in a laboratory frame motion as shown in Figure 4.2.

Other applied forces F perpendicular to B also cause orbital drift, since they are equivalent to an effective electric field $\mathbf{E}_{\text{eff}} = \mathbf{F}/q$. For example, in the presence of a gravitational field g, $\mathbf{E}_{\text{eff}} = (m/q)\mathbf{g}$ and the guiding center drifts with velocity

$$\mathbf{V} = \frac{c}{(q/m)} \frac{\mathbf{q} \times \mathbf{B}}{B^2}. \tag{4.27}$$

Note that V now *does* depend on q and m.

Another important drift is one associated with a magnetic field gradient ∇B. Consider a magnetic field with a longitudinal gradient $\partial B_z/\partial z$, such as in a tapered loop (Figure 4.3). Assuming axial symmetry, so that $\partial/\partial\phi = 0$, the condition $\nabla \cdot \mathbf{B} = 0$ implies

$$\frac{1}{r}\frac{\partial}{\partial r}(rB_r) + \frac{\partial B_z}{\partial z} = 0, \tag{4.28}$$

which integrates to

$$B_r = -\frac{1}{r}\int_0^r r'\frac{\partial B_z}{\partial z}\,\mathrm{d}r' \approx -\frac{r}{2}\frac{\partial B_z}{\partial z}, \tag{4.29}$$

where the variation $\partial B_z/\partial z$ is assumed independent of the radial coordinate r. A particle gyrating about the guiding field line will have a velocity in the azimuthal direction, perpendicular to B_r and, because of the B_r introduced by (4.29), will experience an additional Lorentz force of magnitude

$$F = \frac{q}{c}v_\theta B_r = \frac{\tfrac{1}{2}mv^2}{B_z}\frac{\partial B_z}{\partial z}, \tag{4.30}$$

where we have used equation (4.20) to determine r. Consideration of the directions of the various vector quantities $(\mathbf{B}, \mathbf{r}, \mathbf{v})$ involved shows that $\mathbf{F}$ is in the direction opposite to ∇B_z, so that we may generalize (4.30) to the form

$$\mathbf{F} = -K_\perp \nabla(\ln B), \tag{4.31}$$

where $K_\perp$ is the transverse kinetic energy of the particle. This force causes particles to decelerate in the longitudinal direction as they approach a region of increased field strength. However, as we shall see below, the

Fig. 4.3. Tapered magnetic loop $\mathbf{B} = \mathbf{B}(z)$. The orthogonal coordinate basis (r, ϕ, z) is as shown.

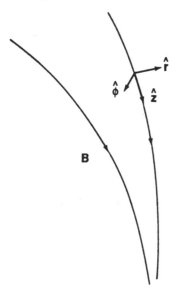

transverse component of energy is correspondingly increased, so as to preserve total energy. The force **F** of equation (4.31) is sometimes called a *mirroring force*.

4.3   Adiabatic invariants

As a charged particle encounters regions of varying magnetic field strength, its motion will be continuously changed in accordance with the Lorentz force law (4.17). However, under appropriate conditions, certain quantities of the motion are conserved; these are termed *adiabatic invariants*. Consider the action integral (round a closed path)

$$J = \oint p_i \, dq_i, \tag{4.32}$$

where p_i is the generalized momentum $= \partial L/\partial \dot{q}_i$, L the Lagrangian, q_i a generalized coordinate, and a dot denotes differentiation with respect to time. Let us consider the variation of J with respect to time:

$$\frac{dJ}{dt} = \oint (p_i \, d\dot{q}_i + \dot{p}_i \, dq_i)$$

$$= \oint \left(\frac{\partial L}{\partial \dot{q}_i} \, d\dot{q}_i + \frac{\partial L}{\partial q_i} \, dq_i \right), \tag{4.33}$$

where we have used Lagrange's equations $\dot{p}_i = \partial L/\partial q_i$ on the second term. In general, the Lagrangian is a function of N coordinates, their time derivatives, and time

$$L = L(q_1, \ldots, q_N, \dot{q}_1, \ldots, \dot{q}_N, t). \tag{4.34}$$

Therefore, taking differentials, we find

$$dL = \sum_{k=1}^{N} \left(\frac{\partial L}{\partial q_k} \, dq_k + \frac{\partial L}{\partial \dot{q}_k} \, d\dot{q}_k \right) + \frac{\partial L}{\partial t} \, dt. \tag{4.35}$$

If we restrict our attention to motions for which all the q_ks and $\dot{q}_k$s are constant, except for $k = i$, then

$$dL = \frac{\partial L}{\partial q_i} \, dq_i + \frac{\partial L}{\partial \dot{q}_i} \, d\dot{q}_i + \frac{\partial L}{\partial t} \, dt, \tag{4.36}$$

and (4.33) may be written

$$\frac{dJ}{dt} = \oint \left(dL - \frac{\partial L}{\partial t} \, dt \right). \tag{4.37}$$

Let us assume that the Lagrangian L is sufficiently constant that the time integral of $\partial L/\partial t$ is essentially zero over the time taken to complete the closed path above. Then

$$\frac{dJ}{dt} = \oint dL = 0; \tag{4.38}$$

i.e., J is a constant. In summary, for sufficiently short time intervals that L does not change appreciably and when all other generalized coordinates and velocities are constant, the action integral $\oint p_i \, dq_i$ is a constant.

As an example, we consider gyration of a charged particle about the Z-axis, so that $q_i = \theta$. Assuming that no external work-performing forces act on the particle (so that L is a constant and the motion can be considered adiabatic), we have

$$\oint p_\theta \, d\theta = \text{constant}. \tag{4.39}$$

The Lagrangian for a particle in an electromagnetic field is [e.g., Symon (1971)]

$$L = \tfrac{1}{2}mv^2 - q\phi + \frac{q}{c} \mathbf{v} \cdot \mathbf{A}, \tag{4.40}$$

where ϕ and $\mathbf{A}$ are the scalar and vector potentials respectively. For the problem at hand, this assumes the form

$$L = \tfrac{1}{2}mr^2\theta^2 - q\phi + \frac{q}{c} r\theta A_\theta, \tag{4.41}$$

so that

$$p_\theta = \frac{\partial L}{\partial \theta} = mr^2\theta + \frac{q}{c} r A_\theta \tag{4.42}$$

and the action integral (4.39) becomes

$$\int_0^{2\pi} \left(mr^2\theta + \frac{q}{c} r A_\theta \right) d\theta = \text{constant},$$

i.e.,

$$2\pi L_z + \frac{q}{c} \int_0^{2\pi} A_\theta(r \, d\theta) = \text{constant}, \tag{4.43}$$

i.e.,

$$2\pi L_z + \frac{q\Phi}{c} = \text{constant}, \tag{4.44}$$

where $L_z = mr^2\dot\theta$ is the Z-component of the angular momentum of the particle and $\Phi = \iint \mathbf{B} \cdot d\mathbf{S} = \iint (\nabla \times \mathbf{A}) \cdot d\mathbf{S} = \oint \mathbf{A} \cdot d\mathbf{l}$ is the magnetic flux contained by the orbit. However, by the relation (4.20),

$$L_z = mvr_B = \frac{q}{c} Br_B{}^2 = \frac{q\Phi}{\pi c}, \tag{4.45}$$

and (4.44) and (4.45) together imply that

$$L_z = \text{constant}, \quad \Phi = \text{constant} \tag{4.46}$$

separately.

The first of these relations is elementary. The second, implying the constancy of magnetic flux threading a gyrating particle's orbit, is not, and constitutes the first adiabatic invariant of the motion; only where $\partial L/\partial t \neq 0$ is Φ not conserved. As a particle moves into a region of larger field strength B_z, its orbit must decrease in radius to preserve Φ, and by the constancy of L_z its azimuthal velocity v_θ must increase. This counteracts the decrease in the particle's v_z velocity due to the mirror force F_z [equation (4.31)], and its total energy is conserved, as must be the case if $\partial L/\partial t$ is essentially zero over one gyration period. Quantitatively, we can use the relations (4.46) to guarantee the constancy of the quantity

$$\frac{\pi L_z{}^2}{\Phi} = \frac{m^2 v_\theta{}^2}{B} = \frac{p_\perp{}^2}{B}, \tag{4.47}$$

where $p_\perp$ is the component of the particle's momentum transverse to the guiding field. If the helical path traced out by the particle has a pitch angle of θ with respect to z, (4.47) may be written (noting that the total momentum p^2 is a constant)

$$\frac{1-\mu^2}{B} = \frac{1-\mu_0{}^2}{B_0}, \tag{4.48}$$

where $\mu = \cos\theta$ and the subscript '0' refers to an (arbitrary) reference point. If electrons are injected at this point with pitch angle cosine μ_0 along a converging magnetic loop, they will have converted all their momentum into transverse momentum when $\mu = 0$, i.e., when

$$1 - \mu_0{}^2 = B_0/B. \tag{4.49}$$

This point represents a *mirror point* for the particles; they cannot pass into a region of greater field strength and so become trapped. The mirror force (4.31) guarantees that such trapped electrons will 'bounce' between the ends of this 'magnetic trap'. Note that, for a prescribed B/B_0, electrons with injected pitch angle cosines $\mu_{inj} > \mu_0$ will escape the trap, since, by (4.48), for these electrons at the mirror point,

$$1 - \mu^2 = \frac{B}{B_0}(1 - \mu_{inj}^2) < \frac{B}{B_0}(1 - \mu_0^2) < 1. \tag{4.50}$$

Thus, the reflected electron distribution will be represented by the originally injected distribution as it passes the injection point, but with a 'loss cone' of missing particles; the half-angle of this cone is $\cos^{-1}\mu_0 = \cos^{-1}[(1 - B_0/B)^{1/2}]$. The stability of such a loss cone will be discussed in Section 7.3.1, in connection with microwave radiation in flares.

Other adiabatic invariants of the motion correspond to taking $q_i = r$ and z, respectively, in the fundamental equation (4.32). However, these results are not of immediate interest in this book, and we refer the interested reader to Chen (1974) for a discussion.

4.4 Collisional dynamics of particle beams

We shall have cause in later chapters to discuss the heating and radiation signatures produced by beams of charged particles. The propagation of such beams is controlled by the electromagnetic forces acting on the particles. These are of several origins – the external electric and magnetic fields discussed in Section 4.2, the response to wave-like electric and magnetic disturbances (Section 3.4), and encounters with other particles, both in the ambient atmosphere and in the beam itself. The physics of the collective interaction of one ensemble of particles with another is quite complex, and we defer a (simplified) discussion to a later section of this chapter. Here we discuss the dynamics of individual particles interacting with a field of charged target particles; see Emslie (1978).

Consider the collision between a test particle of charge ze, mass m, and velocity v with a target particle of charge Ze and mass M. Define the *impact parameter* b of the collision to be the closest distance of approach of an undeflected particle with the same initial velocity (see Figure 4.4). The equations of motion, in (r, θ) coordinates centered on the target particle and with θ measured counterclockwise from the original direction of motion of the test particle (Figure 4.4), are [e.g., Symon (1971)]

$$\ddot{r} = r\dot{\theta}^2 = \frac{Zze^2}{\mu r^2} \tag{4.51}$$

and

$$r^2\dot{\theta} = \text{constant} = -bv, \tag{4.52}$$

where a dot denotes differentiation with respect to time and

$$\mu = \frac{mM}{(m+M)} \qquad (4.53)$$

is the reduced mass of the two-body ensemble. The solution of (4.51) and (4.52) is found by setting $u = 1/r$ and changing the independent variable from t to θ; we then obtain

$$\frac{du^2}{d\theta^2} + u = \frac{-Zze^2}{\mu b^2 v^2}, \qquad (4.54)$$

with solution

$$u = A\cos\theta + B\sin\theta - \frac{Zze^2}{\mu b^2 v^2}. \qquad (4.55)$$

Since initially $\theta = \pi$, $u = 0$, we see that $A = -Zze^2/\mu b^2 v^2$. Also, $v = -\dot{r} = u^{-2}\dot{\theta}\, du/d\theta = -bv(du/d\theta)$, so that $du/d\theta \to (1/b)$ as $\theta \to \pi$, implying $B = 1/b$. Thus, the orbit of the test particle is described by the equation

$$u = \frac{1}{b}\sin\theta - \frac{Zze^2}{\mu b^2 v^2}(1 + \cos\theta). \qquad (4.56)$$

Since the initial energy of the system is positive, closed orbits (negative energy) are impossible, and the orbital path is a hyperbola, as shown in Figure 4.4. The particle goes to infinity ($u = 0$) at values of θ given by

$$\frac{1}{b}\sin\theta = \frac{Zze^2}{\mu b^2 v^2}(1 + \cos\theta).$$

Fig. 4.4. Collision of a test particle of charge ze and mass m with a target particle of charge Ze and mass M, in the frame of the target particle. b is the impact parameter for the collision and the orbit [described by $r(\theta)$] is a hyperbola whose asymptotes intersect at an angle ψ, the scattering angle for the collision.

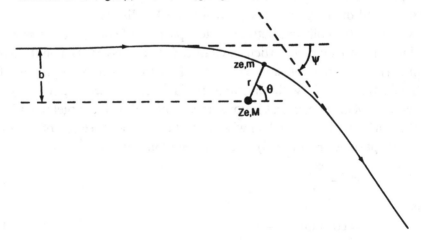

A trivial solution is $\theta = \pi$, corresponding to the incident test particle. The outgoing angle ψ is determined by dividing both sides by $2 \cos^2(\theta/2)$, giving

$$\tan \frac{\psi}{2} = \frac{Zze^2}{\mu b v^2}.$$ (4.57)

For oppositely charged particles ψ is negative (as depicted in Figure 4.4); for particles of the same sign of charge, it is positive. In either case, the angle of deflection (from the original path) is given by (4.57).

The collision itself is elastic as measured in the center of mass frame; in the laboratory frame (solar atmosphere), the test particle transfers energy and momentum to the target particle. In the center-of-mass frame the initial and final velocity vectors of the test particle are

$$\mathbf{v}_i = \frac{M}{m+M}(v, 0), \quad \mathbf{v}_f = \frac{M}{M+m}(v \cos \psi, v \sin \psi),$$

where the components are respectively parallel and perpendicular to the original direction of motion. In the laboratory frame these transform to

$$\mathbf{v}_i = (v, 0), \quad \mathbf{v}_f = \frac{v}{M+m}(M \cos \psi + m, M \sin \psi),$$

corresponding to a change in energy

$$\Delta E = \tfrac{1}{2}mv_f^2 - \tfrac{1}{2}mv_i^2 = \frac{-m^2 M}{(M+m)^2} v^2 (1 - \cos \psi)$$ (4.58)

and a change in parallel velocity

$$\Delta v_\parallel = v_{f,\parallel} - v_{i,\parallel} = \frac{-M}{M+m} v(1 - \cos \psi).$$ (4.59)

These energy and momentum loss rates depend on the impact parameter b through the scattering angle ψ [equation (4.57)]. The evolution of the test particle therefore requires knowledge of the impact parameters associated with successive collisions. Since the number of collisions suffered is very large, this must be done statistically. One can use either a Fokker–Planck approach and treat the problem in terms of the evolution of a velocity distribution function for the electrons (Rosenbluth *et al.*, 1957; Leach and Petrosian, 1981), or perform a Monte Carlo technique and treat the impact parameter b as a random variable with a prescribed statistical distribution [cf. Bai (1982)]. Here we adopt an even simpler approach which turns out to be quite adequate for most modeling purposes. We simply calculate the mean rates of change of E and $v_\parallel$ through the relation

$$\frac{\mathrm{d}}{\mathrm{d}t}(\ ) = 2\pi n v \int \Delta(\ ) \cdot b \, \mathrm{d}b,$$ (4.60)

thus averaging over the impact parameter variable. Here n is the number density of target particles. Using formulas (4.58) and (4.59) in (4.60), and using (4.57) to express θ in terms of b, we derive

$$\frac{dE}{dt} = 4\pi n \frac{m^2 M}{(M+m)^2} v^3 \Xi \, , \qquad (4.61)$$

$$\frac{dv_\parallel}{dt} = -4\pi n \frac{M}{M+m} v^2 \Xi \, , \qquad (4.62)$$

where

$$\Xi = \int_0^{b_0} \frac{b \, db}{1 + \left(\dfrac{\mu b v^2}{Z z e^2}\right)^2} = \frac{1}{2}\left(\frac{Z z e^2}{\mu v^2}\right)^2 \ln\left[1 + \left(\frac{\mu b_0 v^2}{Z z e^2}\right)^2\right]. \qquad (4.63)$$

In deriving this expression, we have placed an upper cutoff b_0 on the allowed range of impact parameters. One reason for this is the breakdown of the Coulomb force law at distances greater than a Debye length (Section 4.1), so that $b_{0,1} = \lambda_D$ would perhaps be a reasonable assumption. However, in the presence of a magnetic field, the range of interaction of an electron is also restricted to values around the gyroradius r_B [equation (4.20)], providing another upper bound $b_{0,2}$. Also, the plasma oscillations of the scattering centers restrict our analysis to times $\tau < \omega_{pe}^{-1}$ [equation (3.57)], so that $b_{0,3} = v\omega_{pe}^{-1}$ is another upper bound. Which of these cutoffs is appropriate will depend on the temperature, density, and magnetic field strength in the target. However, for reasonable solar parameters, we find that the quantity $\mu b_0 v^2 / Z z e^2$ is always so large compared to unity that (i) the unity inside the logarithm can be neglected, and (ii) the value of this Coulomb logarithm remaining, viz.

$$\ln\left[\left(\frac{\mu b_0 v^2}{Z z e^2}\right)^2\right] = 2 \ln \Lambda \qquad (4.64)$$

is fairly insensitive to the precise value of b_0. Typical values of $\ln \Lambda$ for solar conditions are in the range 20–30, and for such values an error of even a factor of 10 in b_0 gives rise to only a 10% error in $\ln \Lambda$. Thus, treating $\ln \Lambda$ as a constant, equations (4.61) and (4.62) become

$$\frac{dE}{dt} = -\frac{2\pi Z^2 z^2 e^4 \ln \Lambda}{E}\left(\frac{m}{M}\right) nv, \qquad (4.65)$$

and

$$\frac{dv_\parallel}{dt} = -\frac{\pi Z^2 z^2 e^4 \ln \Lambda}{E^2}\left(1 + \frac{m}{M}\right) nv^2. \qquad (4.66)$$

Let us now consider the case of electrons incident on an atmosphere of stationary electrons and protons, of equal number densities n. The case of incident protons can be treated similarly (Emslie, 1978). Thus, $m = m_e$, $z = -1$ and we must sum over collisions with electrons ($Z = -1$, $M = m_e$) and protons ($Z = +1$, $M = m_p$). The resulting expressions are

$$\frac{dE}{dt} = \frac{-C}{E} nv \tag{4.67}$$

and

$$\frac{dv_{\parallel}}{dt} = \frac{-3C}{2E^2} nv^2, \tag{4.68}$$

where we have neglected terms of order (m_e/m_p) and set

$$C = 2\pi e^4 \ln \Lambda. \tag{4.69}$$

We see that electron–electron collisions constitute all of the energy loss and two-thirds of the momentum loss. Electron–proton collisions cause no change in energy, but do deflect the electron (elastically) from its pre-collision trajectory. This is seen to be a reflection of the fact that the center-of-mass system (in which the collision is elastic) is identical to that of the target proton, if we neglect terms of order (m_e/m_p). The results (4.67) and (4.68) are easier understood if we express them in terms of the column depth, N,

$$N = \int_0^z n(z') \, dz' \tag{4.70}$$

as the independent variable, so that $(d/dt) = nv_z(d/dN)$. The result (4.67) becomes

$$\frac{dE}{dN} = \frac{-C}{\mu E}, \tag{4.71}$$

where $\mu = (v_z/v)$ is the cosine of the pitch angle of the streaming electron relative to the guiding magnetic field lines (Z-direction). The result (4.68) is better cast in terms of a rate of change of μ; thus, using the relations

$$\frac{d\mu}{\mu} = \frac{dv_z}{v_z} - \frac{1}{2}\frac{dE}{E}, \tag{4.72}$$

(from $\mu = v_z/v$), and $dv_z = \mu \, dv_{\parallel}$, we obtain

$$\frac{d\mu}{dN} = \frac{-C}{E^2}. \tag{4.73}$$

Equations (4.71) and (4.73) have the solution

$$E = E_0 \left(1 - \frac{3CN}{\mu_0 E_0^2} \right)^{1/3}, \tag{4.74}$$

$$\mu = \mu_0 \left(1 - \frac{3CN}{\mu_0 E_0^2} \right)^{1/3}, \tag{4.75}$$

where E_0 and μ_0 are the values at $N=0$ (the point of injection into the target). The electrons stop at a column depth

$$N_{\text{stop}} = \frac{\mu_0}{3C} E_0{}^2 \approx 10^{17}\mu_0[E_0(\text{keV})]^2, \tag{4.76}$$

at which point they also have a mean pitch angle of $90°$ (i.e., are isotropically distributed). The energy lost per unit column depth as a function of column depth N up to this point is given by (4.71), (4.74), and (4.75):

$$\frac{dE}{dN} = -\frac{C}{\mu_0 E_0\left(1 - \dfrac{3CN}{\mu_0 E_0{}^2}\right)^{2/3}}. \tag{4.77}$$

This formula, when combined with a suitable injected energy distribution $F_0(E_0)$, gives the atmospheric heating rate as a function of N (Section 8.1).

Addition of a converging magnetic field gives rise to a further decrease in μ, with no corresponding decrease in E, according to the adiabatic invariant (4.48), which leads to a term

$$\frac{d\mu}{dN} = -\frac{(1-\mu^2)}{2\mu N^*}, \tag{4.78}$$

where $N^*(N) = (d \ln B/dN)^{-1}$. These equations cannot be solved analytically, but approximate solutions have been devised (Chandrashekar and Emslie, 1987).

The formula (4.68) is also useful in assessing the resistance of a plasma to a bulk current flowing through it. According to (4.68), the rate of momentum loss is

$$\frac{dp_\parallel}{dt} = m\frac{dv_\parallel}{dt} = \frac{-3C}{E}n. \tag{4.79}$$

In a steady state this is balanced by the force of an applied electric field E_a, so that

$$-eE_a - \frac{3Cn}{E} = 0. \tag{4.80}$$

Since the current density $j = -nev$, where v now represents a bulk drift, we find that

$$E_a = \frac{3C}{e^2 Ev} j,$$

which is in the form of an Ohm's law [equation (3.35)], with the resistivity η given by

$$\eta = \frac{3C}{e^2 Ev} = \frac{6\pi e^2 \ln \Lambda}{Ev}. \tag{4.81}$$

In a thermal plasma, we may, to order of magnitude, replace E and v by the thermal energy and velocity, respectively, absorbing any inaccuracies into the constant $\ln \Lambda$. Then

$$\eta \approx \frac{6\pi e^2 m^{1/2}}{k^{3/2}} \ln \Lambda \quad T^{-3/2}. \tag{4.82}$$

We note that η is independent of the density n. Substituting numerical values [cf. Spitzer (1962)], we find

$$\eta \approx 8 \times 10^{-8} \ln \Lambda \quad T^{-3/2} \tag{4.83}$$

[cf. equation (3.40) which neglects electron–electron collisions].

4.5 Kinetic theory and collective effects

The interaction of an ensemble of particles with another ensemble is considerably more complicated than the single particle treatment of the previous section, because of the interaction of the particles with other members of their own ensemble – a so-called *collective effect*. Thus, for example, the changing magnetic fields associated with a streaming group of electrons create induced electric fields and affect the evolution of the stream in ways not encountered when considering single particles only.

In cases involving many particles, it is not possible to follow the detailed behaviors of the individual particles; rather, we must treat the problem statistically in terms of the evolution of a distribution function $f(\mathbf{r}, \mathbf{v}, t)$, which defines the number of particles per unit volume and per unit volume of velocity space as a function of position, velocity, and time. The fundamental equation satisfied by f is equation (3.1), the Boltzmann equation; i.e.,

$$\frac{\partial f}{\partial t} + (\mathbf{v} \cdot \nabla) f + \frac{\mathbf{F}}{m} \cdot \frac{\partial f}{\partial \mathbf{v}} = \left(\frac{\partial f}{\partial t} \right)_{\text{coll}}, \tag{4.84}$$

where the left-hand side is the total derivative df/dt, and the right-hand side reflects the removal of particles from the distribution due to collisions. It is the third term on the left-hand side that is responsible for collective behavior, since the principal origin of $\mathbf{F}$ is the electromagnetic force $e(\mathbf{E} + (\mathbf{v}/c) \times \mathbf{B})$, and $\mathbf{E}$ and $\mathbf{B}$ in turn depend on $f(\mathbf{r}, \mathbf{v}, t)$.

We illustrate this collective behavior by two examples. In the first we consider the steady state injection of a beam of electrons into a strongly magnetized collisionless plasma, with $B = B_0 \mathbf{z}$. Since the plasma is collisionless, (4.84) becomes

$$\frac{\partial f}{\partial t} + (\mathbf{v} \cdot \nabla) f - \frac{e}{m} \left(\mathbf{E} + \frac{\mathbf{v} \times \mathbf{B}}{c} \right) \cdot \frac{\partial f}{\partial \mathbf{v}} = 0, \tag{4.85}$$

which is the Vlasov equation. The effect of the magnetic field $\mathbf{B}$ is to cause the

electrons to execute helical paths about the Z-direction (Section 4.2), and we
see that, in the case of a uniform guide field, the $\mathbf{v} \times \mathbf{B}$ force averages to zero
over an orbit. Thus, on timescales substantially larger than the gyration
period ω_B^{-1}, we may write

$$v\frac{\partial f}{\partial z} - \frac{eE}{m}\frac{\partial f}{\partial v} = 0, \tag{4.86}$$

where we have set $\partial f/\partial t = 0$, by the steady state assumption, and recognized
the fact that v and E lie along the Z-direction, so that $f = f(z, v)$. Equation
(4.84) is solved by the method of characteristics, yielding the solution

$$f(z, v) = G(\tfrac{1}{2}mv^2 - e\phi(z)), \tag{4.87}$$

where G is an arbitrary function and ϕ is the electrostatic potential;
$\phi = -\int E\,dz$.

   If $\phi(z)$ were a prescribed function, equation (4.87) would simply reflect the
change in energy of the electrons as they pass through the potential
difference ϕ. However, in this problem, $\phi(z)$ is not defined *a priori*, but
instead results from the injection of the electron beam from the acceleration
site. The introduction of this beam current results in a charge separation,
whose associated electric field leads to the development of an equal and
opposite return current, which is set up in the ambient plasma in order to
replenish the acceleration region. The passage of this return current through
the finite resistivity of the ambient plasma is responsible for the potential
drop, ϕ, through the relation

$$\frac{d\phi}{dz} = -\eta j_{\text{return}} = \eta j_{\text{beam}} = -\eta e \int_0^\infty vf(z, v)\,dv. \tag{4.88}$$

We note that although the streaming electrons may have a high velocity and
be collisionless, the return current constitutes a slow drift of the bulk
ambient plasma and must be treated collisionally.

   Equations (4.87) and (4.88) constitute a set of simultaneous equations for
$f(z, v)$ and $\phi(z)$. Following Knight and Sturrock (1977), we combine
equations (4.87) and (4.88) to find

$$\frac{d\phi}{dz} = -\frac{\eta e}{m} \int_0^\infty G(K - e\phi)\,dK, \tag{4.89}$$

where $K = \tfrac{1}{2}mv^2$ is the kinetic energy of a streaming electron. From the
expression for the total energy

$$U = K - e\phi. \tag{4.90}$$

we derive

$$\frac{d\phi}{d\xi} = \frac{-e}{m} \int_{-e\phi}^\infty G(U)\,dU, \tag{4.91}$$

where $\xi = \int \eta \, dz$ is the resistivity-weighted length along the loop. Differentiation of (4.91) with respect to ξ gives

$$\frac{d^2\phi}{d\xi^2} + \frac{e^2}{m} G(-e\phi) \frac{d\phi}{d\xi} = 0, \tag{4.92}$$

which may be solved for $\phi(\xi)$ given a suitable form for the injected velocity distribution $f(0,v) = G(\frac{1}{2}mv^2)$. Equation (4.90) then gives K, reflecting the slowing of the electrons under the action of the decelerating electric field associated with the return current, and equation (4.87) provides the formal solution to the problem.

The essential *qualitative* result of this analysis is that the kinetic energy K decreases at a rate proportional to the total injected flux [equation (4.91)]. Thus, the individual behavior of the electrons is affected by the number of electrons injected; whereas, in the collisional analysis of Section 4.4, it was determined solely by the structure of the ambient atmosphere. Equation (4.91) represents a collective effect, which may be termed a coordinate–space collective effect, since the collective electric field affecting the behavior of the individual electrons arises from a physical separation of the electrons of the ensemble.

Other collective effects may result from separation of electrons in *velocity* space. To study such velocity–space collective effects, we return to a consideration of plasma oscillations (Section 3.4), but this time from a kinetic viewpoint. Consider a velocity distribution $f(\mathbf{v}, \mathbf{r}, t)$ which is homogeneous in $\mathbf{r}$ and constant in time: $f = f_0(\mathbf{v})$. We introduce a small perturbation $f_1(\mathbf{v}, \mathbf{r}, t)$, and the linearized Vlasov equation (4.84) becomes

$$\frac{\partial f_1}{\partial t} + (\mathbf{v} \cdot \nabla)f_1 - \frac{e}{m} \mathbf{E}_1 \cdot \frac{\partial f_0}{\partial \mathbf{v}} = 0. \tag{4.93}$$

Fourier analysis (cf. Section 3.4) gives

$$-i\omega f_1 + i(\mathbf{k} \cdot \mathbf{v})f_1 = \frac{e}{m} \mathbf{E}_1 \cdot \frac{\partial f_0}{\partial \mathbf{v}}, \tag{4.94}$$

or

$$f_1 = \frac{ie\mathbf{E}_1}{m} \cdot \frac{\partial f_0/\partial \mathbf{v}}{\omega - \mathbf{k} \cdot \mathbf{v}}. \tag{4.95}$$

The linearized form of Poisson's equation is [equation (3.55)]

$$i\mathbf{k} \cdot \mathbf{E}_1 = -4\pi e n_1 = -4\pi e \iiint f_1 \, d^3\mathbf{v}, \tag{4.96}$$

and, since we are looking for electrostatic oscillations, $\nabla \times \mathbf{E} = 0$; i.e.,

$$\mathbf{k} \times \mathbf{E} = 0. \tag{4.97}$$

Thus, eliminating f_1 between (4.95) and (4.96), and setting $\mathbf{E}_1 = \hat{\mathbf{k}} E_1$, we obtain the dispersion relation for the Langmuir waves in the form

$$1 + \frac{4\pi e^2}{mk} \hat{\mathbf{k}} \cdot \iiint \frac{\partial f_0/\partial \mathbf{v}}{\omega - \mathbf{k} \cdot \mathbf{v}} \, d^3 v = 0. \tag{4.98}$$

Let us first simplify this expression by considering only waves propagating in the X-direction ($\hat{\mathbf{k}} = \hat{\mathbf{x}}$), and by introducing the normalized distribution of X-velocities

$$g_0(v_x) = \frac{1}{n_0} \int_{-\infty}^{\infty} \int_{-\infty}^{\infty} f_0(v_x, v_y, v_z) \, dv_y \, dv_z. \tag{4.99}$$

Then equation (4.98) reduces to

$$1 + \frac{\omega_{pe}^2}{k} \int_{-\infty}^{\infty} \frac{\partial g_0/\partial v_x}{\omega - kv_x} \, dv_x = 0, \tag{4.100}$$

where we have used definition (3.75). Integrating by parts, we obtain

$$\omega_{pe}^2 \int_{-\infty}^{\infty} \frac{g_0(v_x) \, dv_x}{(\omega - kv_x)^2} = 1, \tag{4.101}$$

where we have assumed that $g_0(v_x)$ falls off rapidly at $v_x = \pm \infty$ (as, for example, in the case of Maxwellian distribution).

The form of the dispersion relation depends on $g_0(v_x)$. For example, if $g_0(v_x) = \delta(v_x)$, i.e., a cold plasma, the dispersion relation (4.98) becomes

$$\frac{\omega_{pe}^2}{\omega^2} = 1,$$

in agreement with equation (3.75). For a warm plasma at temperature T, the velocity distribution function is

$$g_0(v_x) = \left(\frac{m}{2\pi kT}\right)^{1/2} \exp(-mv_x^2/2kT). \tag{4.102}$$

The dispersion relation in this case is easily evaluated only in the limit when $\omega/k \gg v_{th} = (2kT/m)^{1/2}$, corresponding to long-wavelength waves. An expansion of (4.101) in this case now gives

$$\frac{\omega_{pe}^2}{\omega^2} \int_{-\infty}^{\infty} g_0(v_x) \left(1 + \frac{2kv_x}{\omega} + \frac{3k^2 v_x^2}{\omega^2} + \frac{4k^3 v_x^3}{\omega^3} + \cdots \right) = 1$$

and, keeping the leading terms [note that for a Maxwellian $g_0(v_x)$, the odd terms integrate to zero], we obtain

$$\frac{\omega_{pe}^2}{\omega^2} \left(1 + 3\frac{k^2}{\omega^2} \frac{kT}{m}\right) = 1. \tag{4.103}$$

Since $\mathbf{k}$ is small, $\omega_{pe} \approx \omega$, and in the second term in the brackets we may replace ω by ω_{pe}. Thus,

$$\omega^2 = \omega_{pe}^2 + \frac{3kT}{m}k^2, \tag{4.104}$$

in agreement with the fluid equation (3.77), since $\gamma = 3$ for one-dimensional motion.

Let us now investigate the form of the dispersion relation for the two-stream distribution

$$g_0(v_x) = \frac{1}{2}\left(\delta\left(v_x - \frac{V}{2}\right) + \delta\left(v_x + \frac{V}{2}\right)\right). \tag{4.105}$$

This distribution represents two sets of cold plasmas moving with relative velocity V and has no simple single fluid analog. The dispersion relation (4.101) becomes

$$\omega_{pe}^2\left(\frac{1}{\left(\omega - \frac{kV}{2}\right)^2} + \frac{1}{\left(\omega + \frac{kV}{2}\right)^2}\right) = 2. \tag{4.106}$$

Setting $x = \omega/\omega_{pe}$ and $y = kV/2\omega_{pe}$, we arrive at the form

$$\frac{1}{(x-y)^2} + \frac{1}{(x+y)^2} = 2.$$

This is a quartic in x, which by virtue of the symmetry of the distribution function (4.105), is actually a quadratic in $X = x^2$. With $Y = y^2$ this quadratic can be written

$$X^2 + Y^2 - 2XY = X + Y,$$

with solution

$$X = \frac{(2Y+1) \pm (8Y+1)^{1/2}}{2}. \tag{4.107}$$

Since Y is necessarily positive, both values of X are real. However, if

$$(8Y+1)^{1/2} > 2Y + 1,$$

i.e.,

$$4Y(Y-1) < 0,$$

one of the solutions for X is negative, which implies two imaginary roots for $x = X^{1/2}$. One of these imaginary roots represents a growing wave and the other a damped wave. Such modes are possible only if $0 < Y < 1$, or

$$0 < \frac{kV}{\omega_{pe}} < 2; \tag{4.108}$$

otherwise, all roots for ω are real and represent simple oscillations.

The growth of the low k (long-wavelength) modes represented by (4.107) is known as the *two-stream instability*. The growth rate is a maximum when $Y = \frac{3}{8}$ [by equation (4.107), with the negative sign], and the corresponding ω is

$$\omega = \frac{\omega_{pe}}{2\sqrt{2}} \, i. \tag{4.109}$$

The appearance of an imaginary part in ω is a fundamental result of the collective kinetic approach and has no analog in fluid theory.

Now let us return to the general dispersion relation (4.101). In the case of a general $g_0(v_x)$ it is not obvious how to carry out the integral, because of the presence of singularities $\omega - kv_x = 0$. Physically, these singularities correspond to electrons moving at exactly the phase velocity ω/k of the wave. If we consider $g_0(v_x)$ as the sum of a large number of delta functions (a 'multi-stream' description), then an analysis similar to that above for the double delta function (4.105) shows that a large variety of wave modes is possible; in general two for each delta function present. These modes have phase velocities which are in general close to the velocities of the relevant delta function 'beam'; the fact that $\omega - kv_x$ is close to zero for each mode guarantees that the left-hand side of (4.101) can be made sufficiently large to equal unity, even with an arbitrarily small $g_0(v_x)$, corresponding to the number of particles in the stream. In a 'real' plasma, however, particles are perturbed not into individual monoenergetic streams, but spatially without regard to their velocity, and so these 'streaming modes' will be damped heavily by virtue of the density perturbation and will be smeared out by the different velocities of the perturbed particles. We therefore seek solutions to (4.101) that correspond to the long-timescale asymptotic behavior of physically 'realistic' perturbations. The modes given by (3.57), i.e., pure plasma oscillations, clearly satisfy this requirement, and we ask whether those given by (4.104), i.e., Langmuir waves in a warm plasma, also pass this test. To do so, we must evaluate the imaginary part of ω.

The mathematics leading to ω_i is quite involved, involving the inversion of a Fourier–Laplace transform in the complex plane of the transform variable. We refer the reader to Schmidt (1969) for an excellent treatment of the problem. The result is

$$\omega_i = \frac{\pi}{2} \frac{\omega_{pe}^3}{k^2} \frac{\partial g_0}{\partial v_x} \approx \frac{\pi}{2} \omega_{pe} v_x^2 \frac{\partial g_0}{\partial v_x}, \tag{4.110}$$

so that waves with phase velocity corresponding to parts of the velocity distribution with a negative slope $\partial g_0/\partial v_x$ are damped ($\omega_i < 0$), while those corresponding to parts of the distribution with positive slope $\partial g_0/\partial v_x$ grow with time. This wave damping is known as *Landau damping*; the growth is usually referred to as the two-stream instability because of the special case considered above [equation (4.105) *et seq*; note that a double delta function must possess a positive slope in velocity space].

Physically, Landau damping arises when two groups of particles exchange places in velocity space, one group having gained energy from a wave-associated potential $(-e\phi)$ and the other group having lost the same amount of energy through interaction with the potential at a point in the wavetrain 180° out of phase with the first point. If the distribution has a negative slope in velocity space, this interchange will involve more particles gaining energy then losing it, so that the wave loses energy to the particle distribution as a whole. On the other hand, if the distribution has a positive slope, this interchange of particles results in an energy loss from the particles and so a growth of the wave.

We thus see that wave damping or growth goes hand in hand with a modification to the particle velocity distribution function, so that a proper treatment of the problem must take into account the subsequent evolution of both the particles and waves. If we neglect nonlinear wave–wave interactions, this evolution is described by the so-called *quasi-linear equations*, which, for waves traveling in one dimension only, take the form

$$\frac{dW}{dt} = \pi\omega_{pe}v_x^2 \frac{\partial g}{\partial v_x} W \tag{4.111}$$

and

$$\frac{dg}{dt} = \frac{\pi\omega_{pe}}{nm_e} \frac{\partial}{\partial v_x}\left(v_x W \frac{\partial g}{\partial v_x}\right). \tag{4.112}$$

We note that the growth rate for the wave energy density W (erg cm^{-3} s^{-1} per unit v_x) is twice that for the electric field E_1 [equation (4.110)], since $W \sim E_1^2$. The interplay between these equations is quite subtle; we see that, roughly, concave downward parts of the distribution function g lose particles, while concave upward parts gain them. This follows from the 'group switching' discussion above – particles at a local maximum in the distribution (concave downward) will always be replaced by a lesser number and vice versa for particles at a local minimum (concave upward). If initially $\partial g_0/\partial v_x$ is negative (as, for example, in the case of a Maxwellian), then W decays rapidly (on a time-scale ω_{pe}^{-1}), and so no significant change in g occurs [equation (4.112) with $W \simeq 0$]. If, however, the distribution has an initially positive slope, then W grows rapidly, and only the feedback on $g(v_x)$ through (4.112) limits this growth. The asymptotic solution of (4.111) and (4.112) is a plateau distribution, $g(v_x) = $ constant, with a high level of wave energy caused by the loss of particle energy to the waves. For a double delta function (two-stream) distribution, two-thirds of the initial particle energy ends up in waves (Melrose, 1980). These waves subsequently damp collisionally, at a rate v_c [equation (3.39)], and their energy is thus ultimately used to heat the ambient plasma. However, the spatial and temporal profile of such heating may be substantially modified from the purely collisional results of Section 4.4.

For a two-component plasma, the distribution functions for both the electrons and ions must be considered. The dispersion relations become considerably more complicated, but the essential results are physically plausible. For example, a distribution of electrons streaming with some bulk velocity past a Maxwellian distribution of ions can cause rapid growth of ion-acoustic waves. The threshold for instability is reached when the growth rate exceeds a critical value v_{crit}, given by

$$v_{crit} = c_s \alpha(T_e/T_i), \tag{4.113}$$

where c_s is the ion-sound speed [equation (3.84)]. $\alpha(T_e/T_i)$ is a function of the electron-to-ion temperature ratio, given by Fried and Gould (1961), and well represented by the analytic expression

$$\alpha(T_e/T_i) = \frac{1 + \left(\dfrac{m_i}{m_e}\right)^{1/2}\left(\dfrac{T_e}{T_i}\right)^{3/2}\exp\left[-\dfrac{T_e}{2T_i(1 + k^2\lambda_D^2)}\right]}{(1 + k^2\lambda_D^2)^{1/2}}, \tag{4.114}$$

where λ_D is the Debye length [equation (4.14)]. For a complete treatment of this problem, we refer the reader to Krall and Trivelpiece (1973) and Schmidt (1969). We shall see in Chapter 7 that the flow of a return current [cf. equation (4.88)] can excite such an ion-acoustic instability and may have important implications for models of hard X-ray production in solar flares.

5

Radiative processes in the solar plasma

The bookish theoric

Othello, W. Shakespeare (1564–1616)

In Section 2.2 we discussed the quantum mechanical background for spectral line emission and absorption, which also gave the background for some of the important radiative processes we encounter in solar flares. Radiation is not only an important diagnostic of conditions in flaring atmospheres, but can also be a significant component of the overall energy budget. It is therefore appropriate to consider radiative processes, both atomic and others, in more detail.

Most solar plasmas are highly ionized, which means that generally more than half of all particles present are free electrons. As a consequence, electrons will play the dominant role in most radiative processes. We can treat all these processes classically and derive reasonably good approximate formulae, or we can resort to more sophisticated discussions. It will turn out that while some radiative processes can only be properly understood from a quantum mechanical point of view, others are well suited for the classical treatment.

Classical astrophysics provides for basically four ways in which radiation is created in a stellar plasma, viz.

   (i) scattering of incoming photons by electrons;
   (ii) interaction of electrons with heavier particles;
   (iii) interaction of electrons with magnetic fields; and
   (iv) coherent oscillations of plasma electrons.

To these must be added, under the conditions found in flare plasmas, both relativistic corrections to some of the above processes, as well as new ways of creating radiation, e.g., nuclear γ-ray emission and Compton scattering.

5.1   Scattering

In low-density plasmas, scattering of an external radiation field, being proportional to the electron density n_e, may be more important than the optically thin radiation from the plasma itself, since the latter is proportional to $n_e{}^2$. We shall therefore examine the dependence of scattered radiation intensity on the density. Following Billings (1966) we have an expression for the electric field E^1 generated at a (large) distance r from an accelerated electron

$$E^1 = \frac{e \sin \alpha}{c^2 r} \frac{dv}{dt},$$ (5.1)

where α is the angle between the direction of acceleration (dv/dt) and the line of sight to the electron. If E is the electric field that accelerates the electron ($dv/dt = eE/m_e$), we find $E^1 = e^2 E/mrc^2 \sin \alpha$. If E is oscillatory, so will be E^1, and $(E^1/E)^2$ will give the power in the scattered radiation relative to the incident radiation, so that, the differential cross-section $d\sigma$ for scattering into the solid angle $d\Omega$ is given by

$$\frac{d\sigma}{d\Omega} = r^2 \left(\frac{E^1}{E}\right)^2 = \frac{e^4}{m_e{}^2 c^4} \sin^2 \alpha.$$ (5.2)

For $\alpha = \pi/2$, the electron scattering cross-section is

$$\sigma = \frac{e^4}{m_e{}^2 c^4} = 7.95 \times 10^{-26} \quad \text{cm}^2 \text{ sr}^{-1}.$$

Integrating over the sphere, we find the Thomson cross-section

$$\sigma_e = \frac{8\pi}{3} \sigma = 6.665 \times 10^{-25} \quad \text{cm}^2.$$ (5.3)

Equations (5.2) and (5.3) show that the scattering is independent of frequency. However, this holds only for sufficiently low frequencies where a classical description is adequate. When the frequency v becomes very large so that the energy of the photon, hv, approaches the energy mc^2, quantum effects enter the picture; see Section 5.6. This situation only becomes relevant for electron scattering when hv approaches 0.5 MeV.

Note that from the way in which we derived equation (5.2), using an oscillating electric field, it follows that the differential cross-section pertains to plane-polarized radiation, and in (5.2) we should write $(d\sigma/d\Omega)_{\text{pol}}$. For unpolarized radiation, we derive the equivalent expression by considering such radiation as the independent superposition of two orthogonal, linearly polarized beams. One then finds

$$\left(\frac{d\sigma}{d\Omega}\right)_{\text{unpol}} = \frac{1}{2} \frac{e^4}{m_e{}^2 c^4} (1 + \sin^2 \alpha).$$ (5.4)

If we introduce the angle θ between the incident and scattered waves, equation (5.4) can be written

$$\left(\frac{d\sigma}{d\Omega}\right)_{unpol} = \frac{1}{2}\frac{e^4}{m_e^2 c^4}(1 + \cos^2\theta) \tag{5.5}$$

since $\theta = (\pi/2) - \alpha$.

Such electron-scattered photospheric radiation gives rise to the so-called K-corona and is used to deduce the density of this component of the coronal plasma. A similar procedure may be used to find the density in a flare 'blob' in the corona; see Chapter 9. If we make the assumption of isotropic scattering, we can relate the scattered radiation J to the electron density and the specific intensity. Using equation (5.4), we find that

$$J = \frac{1}{4\pi}n_e I \iint \frac{d\sigma}{d\Omega}\,d\Omega = \tfrac{2}{3}\sigma n_e I. \tag{5.6}$$

Hence, from a measurement of J/I, the density n_e may be determined (see Section 9.2.2).

5.2   Radiation from accelerated free electrons

Two types of radiation are involved here. If the interaction of the electrons with the ions leaves the electrons free, the resulting photons constitute the free–free continuum emission or *Bremsstrahlung*, from the German word for 'braking radiation'. If, on the other hand, the electron–ion collisions lead to the capture of the electrons onto the ion, we have free–bound transitions, and the associated photons form the free–bound continuum.

5.2.1   *Bremsstrahlung*

This emission mechanism may be the most common source of flare radiation. In its nonrelativistic limit, which applies to all optical flare spectra, the process is adequately described in classical terms. However, as we go to higher and higher electron energies (velocities relative to the ions) and refer to observations in the hard X-ray part of the spectrum, quantum corrections come into play.

The most energetic spectral lines of importance in solar physics are the Lyman series of Fe XXVI. The continuum edge for this series lies at $(26)^2 \times 13.6\,\text{eV} = 9.2\,\text{keV}$, so that all radiation above this energy must be in the form of continua, both free–bound and free–free. As the energy increases in this range, the contribution of free–bound emission to the total emission falls off relative to the free–free contribution (Culhane and Acton, 1970), and we find that hard X-ray emission in the deka–keV range is predominantly free–free continuum. We shall see in Chapter 7 that there is considerable interest in the impulsive hard X-ray signature of a solar flare; consequently, we shall now discuss the mechanism of Bremsstrahlung production at some length.

Classical electrodynamics [e.g., Jackson (1962)] predicts that an accelerated electron will emit radiation throughout its acceleration, and Kramers (1923) derived an expression for the radiation rate in free–free transitions as a function of the acceleration of the electron. Using equation (5.1) and remembering that the energy density in the radiation is

$$S = \frac{E^2}{8\pi} + \frac{B^2}{8\pi} = \frac{E^2}{4\pi},$$

(5.7)

we find

$$S = \frac{eE^2}{4\pi} = \frac{e^2 \sin^2 \alpha}{4\pi c^3 r^2} \left(\frac{dv}{dt}\right)^2.$$

The rate of emission, or power, is found by integrating over Σ, the surface of a sphere at radius r

$$P = \int S \, d\Sigma = \frac{2}{3} \frac{e^2}{c^3} \left(\frac{dv}{dt}\right)^2 \quad \text{erg s}^{-1}.$$

(5.8)

With the addition of quantum physics into the analysis, we find that this radiation is composed of discrete quanta (photons) of energy whose characteristics depend on the size and duration of the perturbing force. For relatively distant encounters between electrons and photons, the energy loss of the electron is small and therefore a high-energy quantum cannot be produced. Only for close binary collisions between electrons and protons (or electrons and electrons) can hard X-ray emission result. For such close collisions, a classical description of the electron trajectory itself is not possible, and we must resort to a quantum mechanical description of the collision process, in terms of a cross-section for the particular processes under consideration. In general, Bremsstrahlung cross-sections are functions of the relative velocity of test and target particles both before and after the interaction, of the exit directions of the test particle and photon produced, etc. It is therefore useful to introduce the *differential cross-section* for a process: for example, the quantity $d^2\sigma/dE \, d\Omega$ represents the cross-section, differential in energy E and in the solid angle Ω of the outgoing photon. Note that in the remainder of the discussion, E will denote the energy of the particle, not to be confused with the electric field strength of the radiation.

Cross-sections for the Bremsstrahlung process have been tabulated by Koch and Motz (1959). At high values of the ratio of photon energy ε to electron energy E, the cross-section is highly anisotropic (Gluckstern and Hull, 1953; Tseng and Pratt, 1973). Further, at values of ε (or E) comparable to 511 keV, the rest energy of an electron, relativistic corrections are important. Nevertheless, for the deka–keV energies characteristic of solar hard X-ray bursts (Section 7.2), a convenient formula is the direction-integrated, nonrelativistic, Bethe–Heitler cross-section

$$\sigma_B(\varepsilon, E) = \frac{8\alpha}{3} r_0^2 \frac{m_e c^2}{\varepsilon E} \ln \frac{1 + (1 - \varepsilon/E)^{1/2}}{1 - (1 - \varepsilon/E)^{1/2}}, \qquad (5.9)$$

which is differential in photon energy ε, but includes all possible directions of the outgoing electron and all possible directions and polarizations of the outgoing photon. In the Bethe–Heitler formula (5.9), $r_0 = e^2/m_e c^2$ is the classical electron radius, and $\alpha = 2\pi e^2/hc$ the fine structure constant. Numerically,

$$\sigma_B(\varepsilon, E) = \frac{7.9 \times 10^{-25}}{\varepsilon E} \ln \frac{1 + (1 - \varepsilon/E)^{1/2}}{1 - (1 - \varepsilon/E)^{1/2}} \quad \text{cm}^2 \text{ keV}^{-1}, \qquad (5.10)$$

with ε and E in keV. For $E \leqslant \varepsilon$, σ_B is zero – an obvious result – since an electron of a certain energy cannot be responsible for the emission of a photon of higher energy.

Formulae (5.9) and (5.10) are based on Bremsstrahlung interaction of electrons on stationary protons. However, the solar atmosphere is composed of a wide variety of other atomic species whose abundances relative to hydrogen have been tabulated (Allen, 1973). Since the Bremsstrahlung cross-section, *ceteris paribus*, scales as Z^2, where Z is the atomic number of the scattering ion, we see that we must modify (5.10) to the form

$$\sigma_B = \frac{7.9 \times 10^{-25}\overline{Z^2}}{\varepsilon E} \ln \frac{1 + (1 - \varepsilon/E)^{1/2}}{1 - (1 - \varepsilon/E)^{1/2}}, \qquad (5.11)$$

where the factor $\overline{Z^2}$, the abundance-weighted value of Z^2, is approximately 1.4 for solar conditions [see, e.g., Allen (1973) and Emslie *et al.* (1986a)].

In order to use the observed Bremsstrahlung yield as a diagnostic of the exciting electron population, one must make further assumptions regarding the interaction of the Bremsstrahlung-producing electrons with their environment. Broadly speaking, there are two classes of environment in which such electrons can be found. In the first, the electrons have energies much larger than the mean thermal energy of the background plasma (a 'nonthermal' model) and, in the second, they form part of a relaxed distribution of energetic electrons (a 'thermal' model).

5.2.1.1 *Nonthermal Bremsstrahlung*

Let us calculate the hard X-ray Bremsstrahlung flux $I(\varepsilon)$ (photons $cm^{-2} s^{-1} keV^{-1}$), observed at the Earth, resulting from the injection of a beam of suprathermal energetic electrons with a differential energy spectrum $F(E_0)$ (electrons $cm^{-2} s^{-1} keV^{-1}$), over a flare area S. In determining $I(\varepsilon)$, we must further specify the nature of the target as either 'thin' or 'thick'; in the former case, no significant modification to the injected spectrum occurs, while in the latter the electrons are completely stopped or, more accurately, thermalized in the Bremsstrahlung source. This thermalization may, for example, be due to Coulomb collisions of the electrons with ambient electrons (Section 4.4) or collective interactions with each other (Section 4.5). A thin-target scenario would be applicable to electrons injected outward through the corona, to a population of electrons confined in a tenuous magnetic trap (see Section 4.3), in cases where only part of the target is observed (such as in hard X-ray imaging observations; see Chapter 7), or in cases where part of the target is obscured from the observer by the photospheric limb [e.g., Kane *et al.* (1979a)]. A thin-target scenario would also be applicable to observations with a sampling time considerably shorter than the energy loss time of the electrons in an otherwise thick target. In all other cases, a thick-target scenario is appropriate, and modifications to the injected spectrum $F_{inj}(E_0)$ throughout the target must be considered.

Let us first consider the simpler, thin-target situation. The formula for $I(\varepsilon)$ is

$$I(\varepsilon) = \frac{S \Delta N}{4\pi R^2} \int_{\varepsilon}^{\infty} F(E_0) \sigma_B(\varepsilon, E_0) \, dE_0, \tag{5.12}$$

where $R = 1 \, \text{AU}$, $\Delta N = \int_{\text{source}} n_p(s) \, ds$ is the column density of the source observed, and $n_p(s)$ is the ambient proton density as a function of the distance along the injected electron's path.

For a thick-target situation, (5.12) is still valid, provided we identify $F(E_0)$ not as the injected spectrum, but as the target-averaged flux spectrum $F(E)$. However, in constraining flare models, particularly regarding their particle acceleration properties, it is more useful to have information on the actual injected spectrum $F(E_0)$. The relationship between the target-averaged $F(E)$ and the injected $F(E_0)$ is obtained by considering the energy losses suffered by the Bremsstrahlung-producing electrons in the target. For simple (noncollective) energy loss processes, such as Coulomb collisions on ambient particles, we may represent the energy loss rate through an expression of the form

$$dE/dt = -\sigma_E(E) n_p v(E) E. \tag{5.13}$$

For more complicated processes involving plasma collective effects (Section 4.5), the cross-section for energy loss σ_E will also be a function of other variables (Brown and MacKinnon, 1985; Emslie, 1986).

The number of photons emitted per unit energy, centered on ε, by an electron of initial energy E_0 is

$$m(\varepsilon, E_0) = \int_{t_1(E=E_0)}^{t_2(E=\varepsilon)} n_p(s(t)) \sigma_B(\varepsilon, E(t)) v(E(t)) \, dt. \tag{5.14}$$

If we change the integration variable from t to E using (5.13), we obtain

$$m(\varepsilon, E_0) = \int_{\varepsilon}^{E_0} \frac{\sigma_B(\varepsilon, E) \, dE}{E \sigma_E(E)}, \tag{5.15}$$

an expression which is independent of the density distribution in the target and which, naturally enough, reflects only the ratio of Bremsstrahlung to energy loss cross-sections. For more complicated forms of σ_E [e.g., when reverse current losses or collective plasma effects (Section 4.5) are important], $m(\varepsilon, E_0)$ may also depend on the structure of the target atmosphere and on the form of the injected distribution $F(E_0)$. The Bremsstrahlung flux observed at the Earth is obtained by integrating $m(\varepsilon, E_0)$ over the injected distribution $F(E_0)$ and over the flare area S

$$I(\varepsilon) = \frac{S}{4\pi R^2} \int_{E_0=\varepsilon}^{\infty} F(E_0) m(\varepsilon, E_0) \, dE_0, \tag{5.16}$$

or, using (5.15),

$$I(\varepsilon) = \frac{S}{4\pi R^2} \int_{E_0=\varepsilon}^{\infty} F(E_0) \int_{\varepsilon}^{E_0} \frac{\sigma_B(\varepsilon, E) \, dE}{E \sigma_E(E)} \, dE_0. \tag{5.17}$$

We now consider the case where $\sigma_E(E)$ is given by Coulomb energy losses. Then, from equation (4.67),

$$dE/dt = (-C/E) n_p v, \tag{5.18}$$

so that, from the definition (5.13),

$$\sigma_E(E) = C/E^2, \tag{5.19}$$

where $C = 2\pi e^4 \ln \Lambda$ [equation (4.69)]. Substituting (5.19) in (5.17) we find

$$I(\varepsilon) = \frac{S}{4\pi R^2} \cdot \frac{1}{C} \int_{E_0=\varepsilon}^{\infty} F(E_0) \int_{\varepsilon}^{E_0} E \sigma_B(\varepsilon, E) \, dE \, dE_0. \tag{5.20}$$

Comparing this expression with the thin-target expression (5.12), we see that the effective column density ΔN_{eff} for the thick target is given by

$$\Delta N_{\text{eff}} = \frac{1}{C \sigma_B(\varepsilon, E_0)} \int_{\varepsilon}^{E_0} E \sigma_B(\varepsilon, E) \, dE. \tag{5.21}$$

ΔN_{eff} is thus a function of E_0, behaving roughly like E_0^2, and corresponds approximately to the column density required to stop an electron of injected energy E_0 [equation (4.76)]. If we are observing photons with energy ε, then the $1/E$ dependence of $\sigma_B(\varepsilon, E)$ [equation (5.11)] tells us that the principal contributions to this emission are electrons with energies $E \gtrsim \varepsilon$; this is especially true if $F(E_0)$ is a rapidly decreasing function of E_0 [see equation (5.22) below]. Thus, for targets with an actual column density $\Delta N \ll \Delta N_{\text{eff}}(\varepsilon)$, the thin-target approximation should be good; for $\Delta N \gtrsim N_{\text{eff}}(\varepsilon)$ substantial modification to the injected electron distribution at energies of interest to the Bremsstrahlung process occurs, and we must use the thick-target expression. In the absence of spatial resolution, as was the case for all solar flare hard X-ray observations made before 1980, the thick-target expression is appropriate.

Equations (5.12) and (5.20) are integral equations, relating the injected spectrum $F(E_0)$ (the source function) to the observed hard X-ray spectrum $I(\varepsilon)$ (the data function). The kernel of the integral equation for the thin-target case is the Bremsstrahlung cross-section $\sigma_B(\varepsilon, E_0)$, while for the thick-target case [equation (5.10)] it is an energy-weighted function of σ_B, reflecting the influence of collisional energy losses on the Bremsstrahlung-producing electrons. Considerable literature exists on the properties of such integral equations. For the specific problem under investigation, Brown (1971) showed how equations (5.12) and (5.20) can both be reduced to Abel's integral equation, permitting analytic inversion to yield the source function in terms of the properties of the observed $I(\varepsilon)$. Later papers (e.g., Brown, 1975, 1978) examined the stability of this inversion process, with the conclusion that small errors (e.g., photon counting statistics) in $I(\varepsilon)$ result in large differences between the corresponding inferred source functions. As a result, a wide variety of source functions $F(E_0)$ can reproduce a given $I(\varepsilon)$ within acceptable error limits; i.e., the functions may be far apart in function space but fairly close in 'integral of (function multiplied by kernel) space'. We must therefore be cautious when making statements regarding the form of the injected electron distribution based on observations of $I(\varepsilon)$ alone – the 'smoothing' effect of the kernel on the appropriate integral equation makes a definitive determination of the shape of the source function very difficult [see Craig and Brown (1985)] for a discussion of this problem, and similar ones in other areas of astrophysics].

An alternate way of proceeding is to assume a form for the source function $F(E_0)$ and calculate the resulting Bremsstrahlung spectrum $I(\varepsilon)$. By keeping the assumed form of the source function simple, we can then obtain at least a parametric description of how it varies from event to event. Motivated by the fact that observed hard X-ray spectra are reasonably well fitted by straight lines when plotted on log–log axes (Section 7.2.1), we assume a source function of the form

$$F(E_0) = AE_0^{-\delta}. \tag{5.22}$$

We shall in fact see that this yields the desired power-law hard X-ray spectrum in both thin- and thick-target cases. With (5.22), equations (5.12) and (5.20) then give

$$I_{\text{thin}}(\varepsilon) = \frac{S\Delta N A}{4\pi R^2} \frac{\kappa_{\text{BH}} \overline{Z^2}}{\varepsilon} \int_\varepsilon^\infty E_0^{-(\delta+1)} \ln \frac{1 + (1 - \varepsilon/E_0)^{1/2}}{1 - (1 - \varepsilon/E_0)^{1/2}} \, dE_0 \tag{5.23}$$

and

$$I_{\text{thick}}(\varepsilon) = \frac{SA}{4\pi R^2 C} \frac{\kappa_{\text{BH}} \overline{Z^2}}{\varepsilon} \int_\varepsilon^\infty E_0^{-\delta} \int_\varepsilon^{E_0} \ln \frac{1 + (1 - \varepsilon/E)^{1/2}}{1 - (1 - \varepsilon/E)^{1/2}} \, dE \, dE_0, \tag{5.24}$$

where $\kappa_{\text{BH}} = (8\alpha/3)r_0^2 m_e c^2 = 7.9 \times 10^{-25} \, \text{cm}^2 \, \text{keV}$ is the constant in the Bethe–Heitler cross-section [cf. equations (5.9) and (5.10)]. The thin-target integral can be evaluated by parts:

$$\int_\varepsilon^\infty E_0^{-(\delta+1)} \ln \frac{1 + (1 - \varepsilon/E_0)^{1/2}}{1 - (1 - \varepsilon/E_0)^{1/2}} \, dE_0$$

$$= \left[\frac{E_0^{-\delta}}{\delta} \ln \frac{1 + (1 - \varepsilon/E_0)^{1/2}}{1 - (1 - \varepsilon/E_0)^{1/2}} \right]_\infty^\varepsilon$$

$$+ \int_\varepsilon^\infty \frac{E_0^{-\delta}}{\delta} \frac{d}{dE_0} \left[\ln \frac{1 + (1 - \varepsilon/E_0)^{1/2}}{1 - (1 - \varepsilon/E_0)^{1/2}} \right] dE_0.$$

The first term on the right-hand side vanishes at both limits. The second term reduces to

$$\int_\varepsilon^\infty \frac{E_0^{-\delta}}{\delta} \frac{dE_0}{E_0(1 - \varepsilon/E_0)^{1/2}};$$

whence (5.23) becomes

$$I_{\text{thin}}(\varepsilon) = \frac{S\Delta N A}{4\pi R^2} \frac{\kappa_{\text{BH}} \overline{Z^2}}{\varepsilon} \frac{1}{\delta} \int_\varepsilon^\infty E_0^{-(\delta+1)}(1 - \varepsilon/E_0)^{-1/2} \, dE_0.$$

Setting $x = \varepsilon/E_0$ we find

$$I_{\text{thin}}(\varepsilon) = \frac{S\Delta N A}{4\pi R^2} \kappa_{\text{BH}} \overline{Z^2} \frac{B(\delta, \frac{1}{2})}{\delta} \varepsilon^{-(\delta+1)}, \tag{5.25}$$

where $B(a, b)$ is the standard beta function

$$B(a, b) = \int_0^1 x^{a-1}(1 - x)^{b-1} \, dx; \tag{5.26}$$

$B(a, b)$ can be evaluated using the identity

$$B(a, b) \equiv \frac{\Gamma(a)\Gamma(b)}{\Gamma(a+b)}, \tag{5.27}$$

where $\Gamma(a)$ is the gamma (factorial) function

$$\Gamma(a) = \int_0^\infty e^{-x} x^{a-1} \, dx, \tag{5.28}$$

with the properties

$$\left. \begin{array}{l} \Gamma(a) = (a-1)\Gamma(a-1), \\ \Gamma(1) = 1. \end{array} \right\} \tag{5.29}$$

Equation (5.25) shows that the hard X-ray spectrum is itself a power law

$$I(\varepsilon) = a\varepsilon^{-\gamma}, \tag{5.30}$$

with

$$\gamma = \delta + 1 \tag{5.31}$$

and

$$a = \frac{S\Delta N A}{4\pi R^2} \overline{\kappa_{\mathrm{BH}} Z^2} \frac{B(\gamma - 1. \frac{1}{2})}{(\gamma - 1)}. \tag{5.32}$$

Evaluation of the thick-target expression (5.24) is slightly more involved. First it is useful to reverse the order of integration in the double integral

$$\int_\varepsilon^\infty E_0^{-\delta} \int_\varepsilon^{E_0} \ln \frac{1 + (1 - \varepsilon/E)^{1/2}}{1 - (1 - \varepsilon/E)^{1/2}} \, dE \, dE_0$$

$$= \int_{E=\varepsilon}^\infty \ln \frac{1 + (1 - \varepsilon/E)^{1/2}}{1 - (1 - \varepsilon/E)^{1/2}} \, dE \int_{E_0=E}^\infty E_0^{-\delta} \, dE_0. \tag{5.33}$$

Note that the form of the limits changes, as may be verified by considering the area of integration in (E, E_0) space. Equation (5.24) then becomes

$$I_{\mathrm{thick}}(\varepsilon) = \frac{SA}{4\pi R^2 C} \overline{\kappa_{\mathrm{BH}} Z^2} \frac{1}{(\delta - 1)} \int_\varepsilon^\infty E^{1-\delta} \ln \frac{1 + (1 - \varepsilon/E)^{1/2}}{1 - (1 - \varepsilon/E)^{1/2}} \, dE, \tag{5.34}$$

which may be evaluated exactly as for the thin-target case to yield

$$I_{\mathrm{thick}}(\varepsilon) = \frac{SA}{4\pi R^2 C} \overline{\kappa_{\mathrm{BH}} Z^2} \frac{B(\delta - 2, \frac{1}{2})}{(\delta - 1)(\delta - 2)} \varepsilon^{1-\delta}. \tag{5.35}$$

This is again a power law $I(\varepsilon) = a\varepsilon^{-\gamma}$; in this case, however,

$$\gamma = \delta - 1 \tag{5.36}$$

and

$$a = \frac{SA \overline{\kappa_{\mathrm{BH}} Z^2}}{4\pi R^2 C} \frac{B(\gamma - 1, \frac{1}{2})}{\gamma(\gamma - 1)}. \tag{5.37}$$

We notice that the spectral index of the hard X-ray spectrum is two less (the spectrum is harder) than in the thin-target case (5.31); this is seen, on reflection, to be a consequence of the form of $\sigma_E(E)$ [equation (5.19)], which results in a preferential depletion of low-energy electrons and leads to target-averaged spectrum which is two powers of E harder than the injected spectrum. By comparing equations (5.12) and (5.34), we see that the thick-target hard X-ray yield is in fact that of a thin target with an effective injected flux spectrum

$$F_{\text{eff}}(E_0) = \frac{A}{(\delta - 1)} \left(\frac{E_0^2}{C\Delta N} \right) E_0^{-\delta}, \qquad (5.38)$$

which is two powers of E_0 flatter (harder) than the injected spectrum $F(E_0) = AE_0^{-\delta}$. We also note that the ratio of effective thin-target flux to actual injected flux scales as the ratio of the collisional stopping column density $E_0^2/3C$ [equation (4.76)] to the actual column density ΔN of the source. For low energies F_{eff} is therefore much less than the injected flux, since large fractions of the electrons cannot penetrate beyond their collisional stopping length to the deeper regions of the source.

The formulae (5.31) and (5.32), or (5.36) and (5.37), allow us to infer properties of the injected electron flux, the parameters A and δ, from the observed hard X-ray parameters a and γ, once a target model has been assumed and/or surmised from other observations. Related quantities of interest, particularly in the study of flare energetics, are the total electron flux above some reference energy E_1

$$F_1 = \int_{E_1}^{\infty} AE_0^{-\delta} \, dE_0 = \frac{A}{(\delta - 1)} E_1^{1 - \delta}, \qquad (5.39)$$

and the energy flux above the same reference energy

$$\mathscr{F}_1 = \int_{E_1}^{\infty} AE_0^{-\delta} E_0 \, dE_0 = \frac{A}{(\delta - 2)} E_1^{2 - \delta}. \qquad (5.40)$$

5.2.1.2 *Thermal Bremsstrahlung*

We now consider an alternative model for the Bremsstrahlung source, viz., a hot mass of gas with temperatures T sufficiently large that $kT \approx \varepsilon \approx 10\,\text{keV}$; i.e., $T \approx 10^8\,\text{K}$. Thermal electrons with this energy are capable of producing Bremsstrahlung at X-ray energies. Historically, this was the first model proposed to explain observed solar hard X-ray emission (Chubb *et al.*, 1966). It has the advantage that the Bremsstrahlung-producing electrons do not lose a large fraction of their energy in Coulomb collisions with ambient, cooler electrons as they do in the nonthermal thick-target model. This permits a more energetically efficient Bremsstrahlung process, involving fewer energetic electrons to produce an observed hard X-ray flux. We shall see in Chapter 7 that this energetic advantage removes some celebrated problems involving the interpretation of the hard X-ray signature during the impulsive phase.

Let us then consider a volume V, containing a uniform density n_e of hot electrons with a Maxwellian distribution of speeds corresponding to a temperature T, viz.

$$f(v) = 4\pi \left(\frac{m}{2\pi kT}\right)^{3/2} n_e v^2 \exp(-mv^2/2kT). \tag{5.41}$$

The corresponding energy distribution $f_E(E)$ is

$$f_E(E) = f(v)\,\mathrm{d}v/\mathrm{d}E$$

$$= \frac{2n_e}{\pi^{1/2}(kT)^{3/2}} E^{1/2} \exp(-E/kT) \quad \text{electrons cm}^{-3}\,\text{erg}^{-1}. \tag{5.42}$$

The Bremsstrahlung produced by interaction of these electrons with stationary ambient protons, also of number density n_e, is given by

$$I(\varepsilon) = n_e V \int_\varepsilon^\infty f_E(E) v(E) \sigma_B(\varepsilon, E)\,\mathrm{d}E \tag{5.43}$$

$$= \left(\frac{8}{\pi m_e}\right)^{1/2} \frac{n_e^2 V}{(kT)^{3/2}} \kappa_{BH}\overline{Z^2} \int_\varepsilon^\infty \exp(-E/kT) \ln \frac{1+(1-\varepsilon/E)^{1/2}}{1-(1-\varepsilon/E)^{1/2}}\,\mathrm{d}E$$

$$\text{photons s}^{-1}\,\text{erg}^{-1}, \quad (5.44)$$

where we have used the Bethe–Heitler Bremsstrahlung cross-section (5.10). Integrating by parts, and making the change of variable $E = \varepsilon(1+x)$, we find

$$I(\varepsilon) = \left(\frac{8}{\pi m_e kT}\right)^{1/2} \kappa_{BH}\overline{Z^2}(n_e^2 V)\frac{1}{\varepsilon}\exp(-\varepsilon/kT)g(\varepsilon/kT), \tag{5.45}$$

where

$$g(a) = \int_0^\infty \frac{e^{-ax}\,\mathrm{d}x}{[x(1+x)]^{1/2}} \tag{5.46}$$

is a slowly varying function of order unity. We shall neglect this factor in what follows and write equation (5.45) in the form

$$I(\varepsilon) = D\frac{Q}{\varepsilon T^{1/2}}\exp(-\varepsilon/kT), \tag{5.47}$$

where $D = (8/\pi m_e k)^{1/2}\kappa_{BH}\overline{Z^2} = 5.7 \times 10^{-12}\overline{Z^2}$ cm^3 s^{-1} K$^{1/2}$, and $Q = n_e^2 V$ (or, more generally, $\int n_e^2\,\mathrm{d}V$) is the emission measure of the source; see Section 2.6.

The dependence of $I(\varepsilon)$ on $Q = \int n_e^2 \, \mathrm{d}V$ is a feature common to all optically thin emission involving binary collisions (see Chapter 2). The shape of the Bremsstrahlung spectrum is seen to be of the form $\varepsilon^{-1} \exp(-\varepsilon/kT)$, which is at first sight at variance with observations of a power law $a\varepsilon^{-\gamma}$ [equation (5.30)]. However, this spectrum is only valid for a strictly isothermal source at temperature T. In reality, its proximity to cooler surroundings (e.g., the solar chromosphere) will induce temperature variations throughout the source, in accordance with the physics governing the thermal cooling of the source [see, e.g., Emslie and Brown (1980)]. We use the definition of the differential emission measure distribution, $Q(T)$, equation (2.50),

$$Q(T) = n_e^2 \, \mathrm{d}V/\mathrm{d}T, \tag{5.48}$$

to write the Bremsstrahlung yield from a nonisothermal source as

$$I(\varepsilon) = \frac{D}{\varepsilon} \int_0^\infty \frac{Q(T)}{T^{1/2}} \exp(-\varepsilon/kT) \, \mathrm{d}T, \tag{5.49}$$

which, depending on the form of $Q(T)$, can have a wide variety of spectral shapes. In fact, if we make the change of variable $\tau = 1/kT$, we obtain

$$J(\varepsilon) = \varepsilon I(\varepsilon) = \int_0^\infty e^{-\varepsilon\tau} \bar{Q}(\tau) \, \mathrm{d}\tau, \tag{5.50}$$

where

$$\bar{Q}(\tau) = Dk^{-1/2}\tau^{-3/2}Q\left(\frac{1}{k\tau}\right). \tag{5.51}$$

Equation (5.50) shows that the observed spectrum $I(\varepsilon)$ is related to the Laplace transform of the source function $Q(T)$. Inversion of this equation gives

$$\bar{Q}(\tau) = \mathscr{L}^{-1}[J(\varepsilon); \tau] = \frac{1}{2\pi i} \int_{c-i\infty}^{c+i\infty} e^{\varepsilon\tau} J(\varepsilon) \, \mathrm{d}\varepsilon, \tag{5.52}$$

which shows that any observed $I(\varepsilon)$ can result from a thermal plasma with a suitable $Q(T)$, provided this function is positive everywhere (Brown and Emslie, 1988). For example, the power law already referred to, viz.

$$I(\varepsilon) = a\varepsilon^{-\gamma}, \tag{5.53}$$

requires, by (5.52), that

$$\bar{Q}(\tau) = \frac{a\tau^{\gamma-2}}{\Gamma(\gamma - 1)}. \tag{5.54}$$

Thus, by (5.51)

$$Q(T) = \frac{ak^{1-\gamma}}{D\Gamma(\gamma - 1)} T^{1/2-\gamma}, \tag{5.55}$$

and we see that $Q(T)$ is also a power law with spectral index $(\gamma - \frac{1}{2})$.

Equation (5.49) is of the form of an integral equation for the data function $I(\varepsilon)$ in terms of the source function $Q(T)$ and the kernel $\exp(-\varepsilon/kT)/T^{1/2}$. As discussed in the context of nonthermal Bremsstrahlung above, this kernel has broad 'filtering' properties, leading to the effect that a relatively wide variety of source functions yields very dense functions $I(\varepsilon)$. Thus, within the bounds of statistical data noise, it is quite difficult to define $Q(T)$ closely from spectral data alone. However, it is possible to define the first few integral moments $Q^{(\alpha)}(T) = \int_0^\infty Q(T) T^\alpha \, dT$ in terms of the integral moments $I^{(\beta)}(\varepsilon) = \int_0^\infty I(\varepsilon)\varepsilon^\beta \, d\varepsilon$, and then discuss the types of functions that possess these derived moments. The same technique can be applied to analysis of nonthermal Bremsstrahlung models (Brown, 1978). One must also remember that in all hard X-ray data analysis (both in terms of thermal and nonthermal models) some kind of functional form of $I(\varepsilon)$ must be assumed in order to incorporate the response function of the hard X-ray detector over the finite band width it observes. This usually results in an iterative procedure where a best fit to the observed points is used to model the detector response, thereby better defining the points to be fit, and so on [see Dennis (1982)].

5.2.1.3 *Polarization and directionality*

Up to this point we have considered only the total emission in hard X-rays, without regard to its polarization or directionality characteristics. To investigate these, we must use Bremsstrahlung cross-sections which have not been integrated over the outgoing photon direction or over the direction of the electric field vector of the Bremsstrahlung radiation. The relevant cross-sections may be found in Gluckstern and Hull (1953) and have been summarized by Koch and Motz (1959).

If the energetic electrons responsible for the radiation field have an isotropic velocity distribution at all energies, any polarization or directionality associated with a single Bremsstrahlung collision will average out to zero when integrated over the electron population. We therefore see that any residual polarization and/or directionality of the Bremsstrahlung radiation field is a diagnostic of an anisotropy in the electron velocity distribution and represents a test of flare models. Applicable observational data will be discussed, together with observations of Bremsstrahlung spectra, in Section 7.4.

5.3 Gyrosynchrotron radiation

In the picture of classical electrodynamics an electron experiencing centripetal acceleration as it gyrates about magnetic field lines will emit radiation at the gyrofrequency, given by equation (4.21):

$$\omega_B = \frac{eB}{m_e c} = 1.76 \times 10^7 B. \tag{5.56}$$

The acceleration is given by the Lorentz acceleration

$$\frac{d\mathbf{v}}{dt} = \frac{-e}{mc}(\mathbf{v} \times \mathbf{B}), \tag{5.57}$$

which, when inserted into equation (5.8) leads to an expression for the power radiated by the electron, i.e.,

$$P = \frac{4}{3} \frac{\grave{e}^4 k T B^2}{m_e^3 c^5} \quad \text{erg s}^{-1}. \tag{5.58}$$

In this substitution we have made use of the fact that the relevant, perpendicular, component of v has two degrees of freedom, so that $v^2 = 2kT/m_e$.

The gyroradiation in this classical limit can again be described by equation (5.1), or better, by the vector form of equation (5.1), viz.

$$\mathbf{E} = \frac{e^2}{mc^3} \frac{1}{r^3} \mathbf{r} \times [\mathbf{r} \times (\mathbf{v} \times \mathbf{B})]. \tag{5.59}$$

Equation (5.59) describes in general an elliptically polarized wave, and for values of the magnetic field to be expected in active regions of the solar atmosphere, this radiation is quite strong. However, at the gyrofrequency the extraordinary wave emitted has a vanishing velocity; i.e., the plasma inhibits the escape of the wave from the solar atmosphere. Only when the radiation comes from a level above a certain critical level can the radiation escape. This critical level, forming the *isodiaphanous* surface, is not the level from which radiation of frequencies greater than ω_B can escape, but is the level so located that

$$\frac{\omega_B}{\omega} = 1 - \frac{\omega_{pe}^2}{\omega^2}, \tag{5.60}$$

where ω_{pe} is the local electron plasma frequency [equation (3.75)]

$$\omega_{pe}^2 = \frac{4\pi n_e e^2}{m_e}. \tag{5.61}$$

This isodiaphanous surface is located higher in the solar atmosphere than the level given by $\omega = \omega_B$, provided the magnetic field B decreases with height. Consequently, we will normally not observe gyroradiation at the fundamental frequency. On the other hand, radiation at harmonics of the gyrofrequency can escape. This radiation cannot be treated in the approximation given by equation (5.1), but is given by an expression containing a correction term due to Griem (1964), viz.

$$E = \frac{e^2}{m_e c^3 r^3} [r \times [r \times (v \times B)]] \left(1 + \frac{v \cdot r}{cr}\right)^{-2}. \tag{5.62}$$

The correction term takes into account the effect at the observer of the motion of the electron around the magnetic field lines and is negligible for $\beta \equiv v/c \ll 1$.

When the electron velocities become large, i.e., when β is not small, relativistic effects become important and the radiation – which is then referred to as synchrotron radiation – is no longer amenable to the use of equations built on equations (5.1) or (5.62). Schwinger (1949) was concerned with the power in the radiation and found that instead of equation (5.58), P is given by

$$P = \frac{2}{3} \frac{e^4 v^2 B^2}{m_e^2 c^5} (1 - \beta^2)^{-2}. \tag{5.63}$$

As β increases one finds that the radiation becomes highly concentrated in a cone whose angle is $\theta = 1 - \beta^2$ and which points in the direction of v. An observer in the orbital plane of the gyrating electron will see a train of short, highly polarized radiation pulses as the cone sweeps across the observer, the pulses repeating with the gyrofrequency ω_B. A Fourier analysis of the pulsed spectrum reveals a continuous spectrum whose intensity increases with frequency up to a critical frequency ω_c, above which the intensity then quickly falls off. This critical frequency is given by

$$\omega_c = \tfrac{3}{2}\omega_B (1 - \beta^2)^{-3/2}, \tag{5.64}$$

an expression showing that for energetic electrons we can expect synchrotron radiation to play a role over a large part of the astrophysically important spectrum.

The more thorough treatments of gyrosynchrotron radiation quickly become very complicated, even in mildly relativistic cases. The general expression for the emission and absorption coefficients have been worked out by Ginzburg and Syrovatskii (1965), Melrose (1980), and others. To find expressions that are useful for analyzing flare conditions several approximations are available. Distinctions must be made between relativistic versus nonrelativistic electrons and between thermal and nonthermal electrons. Dulk and Marsh (1982) argue that observed impulsive microwave emission from flares comes from electrons with energies between 10 keV and 1 MeV. At these energies neither of the simplifications valid in the nonrelativistic or highly relativistic limits is applicable, and special procedures are necessary.

We have seen that nonrelativistic electrons produce gyromagnetic radiation which is concentrated at the electron gyrofrequency ω_B and its first harmonics $\omega/\omega_B = 2, 3$ (Bekefi, 1966). On the other hand, highly relativistic electrons emit gyromagnetic radiation at very high harmonics ($\omega/\omega_B \gg 100$). The electron energies associated with the gyromagnetic radiation identified as the flare microwave emission by Dulk and Marsh (1982) correspond to harmonics in the range $10 \ll \omega/\omega_B < 100$ (see Section 7.3.1).

In the relatively simple case of a nonrelativistic Maxwellian electron distribution (thermal plasma), Zheleznyakov (1970) (see also Melrose, 1980) showed that the absorption coefficient, α_ω, for gyromagnetic radiation of frequency ω which is a function of the harmonic number, s, the angle θ between the wave-normal direction, and the magnetic field, can be written

$$\alpha_\omega{}^\pi(s, \theta) = \left(\frac{\pi}{2}\right)^{1/2} \frac{\omega_p{}^2 A^\pi(s, \omega, \theta)}{\omega n_\pi{}^2 \left(\dfrac{kT_e}{mc^2}\right)^{1/2} \cos\theta \, \dfrac{\partial(\omega n_\pi)}{\partial\omega}}$$

$$\times \exp\left[-\frac{(\omega - s\omega_B)^2}{2\omega^2 n_\pi{}^2 \dfrac{kT_e}{mc^2}\cos^2\theta}\right]. \tag{5.65}$$

Since we have both an ordinary and extraordinary magneto-ionic mode, the super, or subscript π indicates these modes; i.e., $\pi = +1$ is the o-mode, $\pi = -1$ the eo-mode, n_π is the refractive index of the mode in question. The function $A^\pi(s, \omega, \theta)$ contains the polarization information and is sufficiently complicated to render the formulae less than useful for flare-plasma diagnostics. However, since $\alpha_\omega{}^\pi(s, \theta)$ is peaked sharply around integral values of ω/ω_B, one may work with a mean absorption coefficient relating to the sth harmonic, viz.

$$\alpha^\pi(s, \theta) = \int_{-\infty}^{\infty} \frac{1}{\omega_B} \alpha_\omega{}^\pi(s, \theta) \, d\omega. \tag{5.66}$$

Only by making serious simplifications, such as reducing the power series expansion for the modified Bessel function appearing in $A^\pi(s, \omega, \theta)$ to its leading term, could Zheleznyakov (1970) and Dulk et al. (1979) develop sufficiently simple expressions, viz.,

$$\alpha^\pi = \frac{\pi\omega_p{}^2}{\omega_B}\left[1 - 2\frac{kT_e}{mc^2}\cos 2\theta\left(1 - \frac{9}{4}\frac{\omega_p{}^2}{s^2\omega_B{}^2}\right)\right] A^\pi(s, \omega_o, \theta) \tag{5.67}$$

and

$$A^\pi(s, \omega_o, \theta) = \frac{e^{-\lambda}}{4}\frac{s^2}{s!}\left(\frac{\lambda}{2}\right)^{s-1}\left(1 - \frac{\pi\sin^2\theta}{2s\cos\theta}\right)(1 - \pi\cos\theta)^2$$

$$- (1 - \pi\cos\theta)\sin^2\theta\left[\frac{1}{s} + 2\pi\frac{kT}{mc^2}\cos\theta\right]. \tag{5.68}$$

where

$$\lambda = s^2 \frac{kT}{mc^2} n^2 \sin^2 \theta \left(1 - 2 \frac{kT}{mc^2} n^4 \cos^2 \theta \right)$$

and

$$n \equiv \left(1 - \frac{\omega_p^2}{s^2 \omega_B^2} \right)^{1/2}.$$

In flare plasmas we may expect relativistic effects to invalidate the use of the above formulae, and Trubnikov (1958) looked at such effects for the special case of $\theta = \pi/2$. The analysis showed that an important difference occurs in the value of the expected line width, which may be quite small in the nonrelativistic case [as a matter of fact equal to zero for $\theta = \pi/2$ according to the formula (5.65)]. When relativistic effects are taken into account, one finds a significant spread in the gyrofrequencies, and the peak in the absorption at the sth harmonic occurs at frequencies given by $\omega/\omega_B < s$ (Trubnikov, 1958). Dulk et al. (1979) and Dulk and Marsh (1982) derived approximate formulae useful for analysis of flare emission spectra. They gave the value of the peak frequency, ω_{max}, and peak flux density, S_{max}, for the mildly relativistic case when the electron distribution was nearly Maxwellian, i.e., for quasi-thermal and nonthermal electrons. When the optical depth $\tau = \alpha L$ is large (where L is a characteristic dimension of the emitting plasma), we have

$$\omega_{max} = \begin{cases} \text{constant} \times BT^{1/2}, & \text{thermal,} \\ \text{constant} \times (nL)^{1/3} B^{2/3}, & \text{nonthermal,} \end{cases} \tag{5.69}$$

and

$$S_{max} = \text{constant} \times \omega^2 TL^2. \tag{5.70}$$

Thus, by observing S_{max} an estimate may be made for the size of the emitting plasma if an independent method is available to determine the temperature, e.g., by X-ray observations. Similarly, from equation (5.69) the observation of ω_{max} may lead to an estimate of the magnetic field strength in the source region. We return to these matters in Section 7.3.

5.4   Coherent oscillations

In Section 3.4.1 we analyzed how electrostatic forces in a plasma may set up coherent oscillations of the electrons. The motions could be described as simple harmonic oscillations, and the period given by the electron plasma frequency, equation (3.75):

$$\omega_{pe} = \left(\frac{4\pi n_e e^2}{m_e} \right)^{1/2}.$$

The nature of the plasma oscillations can be studied with equation (3.57) using the fact that in such waves **E** and **j** are parallel to the direction of propagation of the waves. These electrostatic oscillations, therefore, are longitudinal waves, and coupling to transverse electromagnetic waves is necessary for a radiation mechanism to become viable. If such coupling takes place, radiation at the plasma frequency will be generated and several types of flare emission have been identified with such radiation (e.g., radio Type III bursts; Section 7.6.2).

5.5   Čerenkov radiation

This emission process – discovered by Čerenkov in 1934 and first studied theoretically by Tamm and Frank – may be considered a Bremsstrahlung, but we will treat it separately because of the special conditions under which it operates. In an isotropic medium the wave fronts from a continuously emitting source at rest will be concentric spheres spreading out in all directions. If the source moves, the Doppler effect will cause the spherical wave fronts to have their centers on a line traced out by the source with the source at any one time being the center of an emitted front. The spherical wave fronts will be more compressed together in the direction of motion than in the opposite direction, but the wave fronts will not intersect as long as the velocity is subsonic. However, if the source moves through the medium with a speed v greater than the phase velocity v_s of the wave in the medium, the spherical wave fronts will intersect, because in that case newly emitted fronts will overtake previously emitted ones. The result is that the overall wave front envelope takes the shape of a cone with the moving source at its apex; see Figure 5.1.

Fig. 5.1. Čerenkov radiation produced by a particle of speed v traveling through a medium where the group velocity of light v_s is smaller than v.

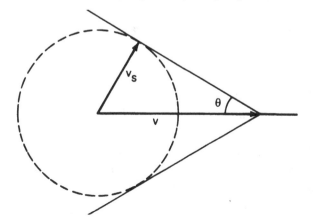

In the case of Čerenkov radiation, very high-energy-charged particles penetrate a plasma in which the velocity of light is smaller than v. The relevant group velocity for light is $v_s = c/n$, where n is the refractive index, and Čerenkov radiation occurs when $v > c/n$. The charged particles are heavily braked by the plasma, and the resulting radiation is emitted in the described cone, whose surface makes an angle θ with the direction of motion, given by

$$\sin \theta = \frac{v_s}{v} = \frac{c}{nv}. \tag{5.71}$$

The emission is restricted to a narrow frequency range near the plasma frequency, and it has been invoked to explain some types of radio frequency emission from flares.

5.6   The inverse Compton effect

In Section 5.1 we discussed scattering of photons by electrons in the classical case of Thomson scattering where the photon energies are small compared to mc^2 and where quantum effects are neglected. In this case the incident photon energy, hv_1, is equal to the scattered photon energy, hv_2; i.e., we have what is called coherent or elastic scattering.

Compton scattering pertains to the case where quantum effects become important. Since the incident photon has a momentum $hv_1/c = h/\lambda_1$, the scattering is no longer elastic. The recoil of the electron (or any other charged particle) leads to the condition $hv_1 \neq hv_2$. If the vector momentum of the photons changes through an angle θ, it can be shown that the scattered photon energy is given by

$$hv_2 = \frac{hv_1}{1 + \frac{hv_1}{mc^2}(1 - \cos \theta)}. \tag{5.72}$$

Equation (5.72) is often expressed in terms of wavelength:

$$\lambda_2 = \lambda_1 + \frac{h}{mc}(1 - \cos \theta). \tag{5.73}$$

The difference $\Delta\lambda = \lambda_2 - \lambda_1$ is the Compton shift, and the ratio $h/mc \equiv \lambda_c$ is referred to as the *Compton wavelength*, which for electrons is $h/m_e c = 0.024$ Å.

Equation (5.72) shows that Compton scattering reduces the energy of the photon. However, a process of interest in solar flare plasmas, referred to as the *inverse Compton effect*, can lead to a significant *increase* in the photon energy, i.e., $hv_2 \gg hv_1$. This can happen whenever the moving electron has a large kinetic energy compared to the incident photon, in which case energy is transferred from the electron to the photon. A rigorous treatment of this situation is outside the scope of this book; however, the following simple discussion outlines the salient features of the process. Consider the head-on collision of a photon of energy $h\nu_1$, with an electron of velocity v, and corresponding relativistic Lorentz factor $\gamma = (1 - v^2/c^2)^{-1/2}$. In the rest frame of the electrons which moves to the right (say) with velocity v, the photon has an energy $\gamma h\nu_1$ and a momentum to the left and, after scattering, will have a similar energy but a momentum to the right. If we now transform this scattered energy back into the laboratory frame, we find that the photon energy after scattering is given approximately by $\gamma(\gamma h\nu_1) = \gamma^2 h\nu_1$; i.e., the inverse Compton effect can transfer a low-energy photon to a high-energy photon with an energy gain of a factor γ^2. Hence, very energetic radiation can be produced if sufficiently energetic ($\gamma \gg 1$) electrons are present. However, if the photon energy in the rest frame of the electron becomes too high, a quantum effect on the scattering cross-section will limit the effectiveness of the inverse Compton effect. In such cases the differential cross-section for unpolarized radiation is not given by equation (5.5). A quantum electrodynamic treatment (Heitler, 1954) shows that the Klein–Nishina formula must be used:

$$\frac{d\sigma}{d\Omega} = \frac{1}{2}\frac{e^4}{m_e^2 c^4}\left(\frac{\nu_2}{\nu_1}\right)^2\left(\frac{\nu_1^2 + \nu_2^2}{\nu_1 \nu_2} - \sin^2\theta\right). \tag{5.74}$$

The corresponding total cross-sections σ are (Heitler, 1954)

$$\sigma = \sigma_e\left(1 - \frac{2\gamma h\nu_1}{m_e c^2}\right), \tag{5.74'}$$

valid when $\gamma h\nu_1 \ll mc^2$ (and $h\nu_2 \approx \gamma^2 h\nu_1$, as above), and

$$\sigma = \tfrac{3}{8}\sigma_e\left(\frac{m_e c^2}{\gamma h\nu_1}\right)\left[\ln\left(\frac{2\gamma h\nu_1}{m_e c^2}\right) + \frac{1}{2}\right], \tag{5.74''}$$

valid when $\gamma h\nu_1 \gg mc^2$ (in which case $h\nu_2 \approx \gamma m_e c^2$).

For $h\nu_1 = h\nu_2$, i.e., an elastic collision, equation (5.74) reduces to the classical expression, equation (5.4). We see from (5.74') and (5.74'') that the main effect of the quantum mechanical treatment on the cross-section is to reduce it from its classical value when the photon energy becomes large, which means that Compton scattering becomes less efficient at high energies.

5.7 Nuclear processes

There is a host of processes involving changes in atomic nuclei that lead to the emission of photons, generally of very high energy, i.e., in the hard X-ray and γ-ray domain of the electromagnetic spectrum. Here we describe some of these processes. An excellent review on nuclear processes in astrophysical plasmas may be found in Ramaty *et al.* (1975). Observations of solar flare γ-rays will be discussed in Section 7.4.

5.7.1   *Decay of radioactive nuclei and excited stable nuclei*

Stable nuclei in their normal ground state can be excited to higher energy levels by collisions with particles or photons. Radioactive nuclei are initially in excited states. In both cases the nuclei will decay from the excited state with energy E_U to a lower state with energy E_L under the emission of a γ-ray photon of energy $h\nu$ and by gaining an amount of energy $K = \frac{1}{2}mv^2$ due to the recoil motion of the nucleus,

$$E_U - E_L = h\nu + \tfrac{1}{2}mv^2.$$

The momentum $p = mv = h\nu/c$ will be conserved, and the energy may be written $E_k = p^2/2m = h^2v^2/2mc^2$, which leads to the following expression for the difference in energy between the upper and lower state of the nucleus:

$$E_U - E_L = h\nu + \frac{h^2v^2}{2mc^2} = h\nu\left(1 + \frac{h\nu}{2mc^2}\right). \tag{5.75}$$

The term $h\nu/2mc^2$ may be considered a correction term to the normally used expression $h\nu = E_U - E_L$.

Under certain flare conditions we may expect γ-rays whose frequency is given by equation (5.75) and the correction term may be important. In the theory of solids, recoilless emission of γ-rays is associated with the so-called Mössbauer effect.

5.7.2   *Positron annihilation*

The free annihilation of positrons with electrons leads to the formation of a γ-ray line with $h\nu = 0.511\,\text{MeV}$. Because of Doppler broadening, when the positrons are relativistic, the annihilation line appears as a γ-ray continuum. For temperatures greater than about $10^6\,\text{K}$, free annihilation is the dominant process by which positrons disappear. However, for lower temperatures, especially for $T > 10^5\,\text{K}$, formation of positronium, followed by positronium decay, dominates the fate of the positrons. Positronium is a strange atom, analogous to hydrogen, with an electron and a positron, e^+, moving in Bohr orbits around their center of mass. The lowest orbit provides for an S state, either a ^1S or a ^3S state, depending on whether the electron spin and the positron spin are oppositely directed or are parallel. The threshold energy ΔE for the formation of positronium, P_s, is 6.8 eV, i.e.,

$$e^+ + e^- \rightarrow P_s + 6.8 \, \text{eV}. \tag{5.76}$$

Positronium in a ^1S state decays under the formation of two γ-ray photons, each of energy 0.511 MeV. On the other hand, positronium in a ^3S state – which is metastable with lifetime 1.4×10^{-7} s or 1200 times longer than for the ^1S state – decays with the formation of three γ-ray photons, each with energy less than 0.511 MeV.

The 511-keV line is referred to as a delayed γ-ray line, since the emission of the line extends over a time interval longer than the production time of the radioactive positron emitter.

If the plasma is more than about 50% ionized, i.e., $n_e/n_{\text{neutral}} > 0.5$, the positrons tend to thermalize before they form positronium by charge exchange

$$e^+ + \text{atom} \rightarrow P_s + \text{atom}^+ + \Delta E, \tag{5.77}$$

where ΔE, the difference in energy in the two systems, can be positive or negative. The charge exchange [equation (5.77)] cross-section has a maximum for $E \simeq 15-20$ eV. If the energy of the plasma electrons, E_e, is greater than 6.8 eV, the γ-ray line at 0.511 MeV that is emitted when the positronium decays will be broadened. For $E_e < 6.8$ eV, the 0.511-MeV line is nearly sharp.

If the plasma is less than about 50% ionized, the positrons form positronium, by equation (5.84), before their energy has decreased to 15–20 eV, and the annihilation rate is determined by charge exchange. Ramaty *et al.* (1975) showed that in the case of decay from the ^1S state the Doppler broadening of the 0.511-MeV line is determined by the kinetic energy of the positronium atom, provided the decay occurs before the positronium has thermalized. The broadening of the 0.511-MeV line and the line/continuum ratio can be used as diagnostic tools in the study of solar flare plasmas.

5.7.3 *Nuclear interaction (capture, inelastic scattering, and spallation)*

Certain γ-ray lines are produced in solar flare plasmas by so-called prompt, discrete processes due to nuclear interactions between high-energy particles (protons, neutrons, α-particles) and the ambient atoms (ions) of the plasma. The inelastic scattering processes can be written as

$$^A_Z\text{N} + {}^1_1\text{p} \rightarrow {}^A_Z\text{N}^* + {}^1_1\text{p}' \tag{5.78}$$

after which the excited nucleus N* reverts to the ground state (or to a lower, still excited state) with the emission of a γ-ray line, viz.

$$\substack{A\\Z}N^* \rightarrow \substack{A\\Z}N + \gamma \tag{5.79}$$

or

$$\substack{A\\Z}N^*(U) \rightarrow \substack{A\\Z}N^*(L) + \gamma'. \tag{5.80}$$

In the above nomenclature A is the atomic weight, Z the atomic number, and p a proton. Equation (5.80) is often written in the form

$$\substack{A\\Z}N(p, p')\substack{A\\Z}N^*, \tag{5.81}$$

or in shorthand, combining the essential information of equations (5.78) and (5.79)

$$\substack{A\\Z}N(p, p\gamma). \tag{5.82}$$

Similarly, inelastic scattering involving an α-particle (He nucleus) and leading to γ-ray line emission is written

$$\substack{A\\Z}N(\alpha, \alpha\gamma) \tag{5.83}$$

and with a neutron involved

$$\substack{A\\Z}N(n, n\gamma). \tag{5.84}$$

Interaction processes different from inelastic scattering but, leading like these to prompt γ-ray emission, comprise reactions like (α, n), (α, p), and (p, n). An important example from solar physics is the reaction

$$\substack{4\\2}He(\alpha, n)\,^7Be^*, \tag{5.85}$$

followed by $^7Be^* \rightarrow {}^7Be + \gamma$, $h\nu = 0.431\,\text{MeV}$.

In a process referred to as *delayed*, a neutron, produced in an earlier spallation reaction (cf. Table I of Ramaty *et al.*, 1975), is captured by the nucleus which is raised to an excited state from which it decays with the emission of a γ-ray photon, viz.

$$\substack{A\\Z}N + \substack{1\\0}n \rightarrow {}^{A+1}_{\ \ Z}N^* \rightarrow {}^{A+1}_{\ \ Z}N + \gamma. \tag{5.86}$$

In solar flares, an example of this process is deuterium formation: $\substack{1\\1}H(n, \gamma)\substack{2\\1}H$ with γ-ray energy $h\nu = 2.223\,\text{MeV}$. Although the capture process is prompt, the cross-section is a decreasing function of the neutron energy. Thus, the line emission is usually significantly delayed in order to allow the neutrons to thermalize before they are captured. In addition, the low energy of the captured neutrons means that the emitted line is very sharp.

Sometimes the collision of a proton or an α-particle with a heavy nucleus will cause the nucleus to break up into lighter fragments that are left in excited states. This process is referred to as spallation, and examples are written $(p, 2p)$, $(p, 2pn)$, $(p, p\alpha)$, etc.

Important examples also found in solar flares are the reactions

$$^{16}O(p, \alpha)^{12}C^* \rightarrow h\nu = 4.44\, \text{MeV} \tag{5.87}$$

and

$$^{20}Ne(p, p\alpha)^{16}O^* \rightarrow h\nu = 6.13\, \text{MeV}. \tag{5.88}$$

5.7.4 π^0 *decay*

A number of mesons are produced in the many possible reactions taking place between high-energy particles. Many of the mesons $(K^0, K^+, K^-, \Lambda, \ldots)$ decay into π mesons, of which the π^0 is of special interest in flare physics. The π^0 meson decays in 10^{-16} s into two γ-rays

$$\pi^0 \rightarrow \gamma + \gamma, \tag{5.89}$$

whose individual energy is $h\nu = 68\, \text{MeV}$ in the rest frame of the pion. Because of the random energies of the individual emitting pions, the observed spectrum will be a broad continuum with a peak near 100 MeV, and no discrete lines are visible [e.g., Kniffen *et al.* (1977); Murphy and Ramaty (1984)].

6

Pre-flare conditions

Guess if you can, choose if you dare

Heraclius, P. Corneille (1600–84)

Flares do not occur just anywhere on the Sun, but only in parts of the atmosphere where conditions are right; namely, in certain active regions. The obvious question, and one which is especially interesting to prospective flare predictors, is: What makes conditions right? Also, we need to ask what characterizes the conditions in these active regions and how they differ from conditions in the rest of the solar atmosphere where flares are extremely unlikely to occur.

6.1 The concept of an active region

One normally refers to an active region as any part of the solar atmosphere that exhibits excess Hα emission. This is a crude observational definition, dictated by the fact that, historically, most pertinent observations have been done in the light of the Hα line. It furnishes a convenient means of observing these active regions where flares may occur. Not all active regions produce observable flares, and certain specific pre-flare conditions are obviously necessary for this to happen.

Both from a theoretical point of view (in order to understand the flare phenomenon) and a practical point of view (in order to be able to forecast where and when a flare will occur), a closer study of pre-flare conditions in active regions is essential. These pre-flare conditions may be characterized by differences from quiet-Sun conditions in the magnetic field that pervades the plasma (the magnetic field is always stronger than in the quiet surroundings) and/or in the dynamic state, i.e., the velocity of the plasma, as well as in its thermodynamics, e.g., temperature and density. The fact that most flares occur in active regions indicates that the pre-flare conditions include (at least) increased values of magnetic field and temperature relative to the quiet solar atmosphere. During the past several years, evidence has also been mounting that the dynamics of the pre-flare plasma is essential in setting up the right conditions. Certain motions of the plasma and their interaction with the pre-existing magnetic field, leading to shears and stresses, seem to play a dominant role.

6.1.1 *Active-region nomenclature*

An active region on the Sun should not be thought of as a two-dimensional area; it extends from deep photospheric levels (or below) high into the transition region and the corona. Historically, the photospheric part of an active region has been observed in white light as bright patches – especially visible toward the solar limb – that are called *faculae* or, more specifically, photospheric faculae to emphasize their location. Also historically, pictures of the Sun in the light of strong spectral lines – like Hα or the Ca II K line – which show similar bright areas reveal chromospheric faculae, since the core of such strong lines is formed in the chromosphere. The French designation is *plage faculaire*; hence, the word 'plage' has been adopted in astrophysical terminology as an alternative name for a chromospheric facula and sometimes for the whole active-region complex. Using soft X-rays, we can probe the active region at low coronal heights, and the observed bright emission (the X-ray plage) may be referred to as the coronal part of the plage, which also may be observed in the radio frequency spectrum as the radio plage. At longer radio waves (meter wavelengths) – which originate in the outer corona – we probe the active region to great distances from the photosphere. On the other hand, the shortest radio waves (centimeter and millimeter wavelengths) originate in the lower corona and the chromosphere and provide us an independent means of exploring this region of the solar atmosphere [see, e.g., Kundu and Alissandrakis (1984)].

6.1.2 *The fundamental loop structure*

The advent of space observations with high spatial resolution, particularly the Skylab observations, revealed that the loop is the fundamental structure element of active-region plasmas. It has certainly been known for some time that solar activity manifestations often take the form of loops (such as post-flare loops and loop structures in the emission line corona, e.g., the Fe XIV 5303-Å line), but it was above all the soft X-ray observations from Skylab that showed the ubiquitous nature of loops in this solar atmospheric plasma. We find that the active solar plasma is, to a large extent, locked up in loops that exist on scales from the smallest bipolar loops emerging from below the photosphere to enormous coronal loops and arches spanning distances of a solar radius or more; see Figure 6.1. Apart from direct Zeeman effect measurements, this plethora of loops of all sizes in and around active regions furnishes the best indication of the overwhelming importance of magnetic fields in solar activity research.

We note that it is the 'frozen-in' property of highly conducting plasmas [equation (3.98)] that allows us to draw these conclusions as to the shape of the magnetic field from observations of the spatial structure of emission. The deduced loop structure of the magnetic field has formed the basis for a variety of flare models involving both isolated [e.g., Spicer, 1977)] and

Fig. 6.1. A large prominence stabilized by loop-shaped magnetic fields of varying dimensions (courtesy M. J. Martres, Observatoire de Paris, Section de Meudon).

colliding (Heyvaerts *et al.*, 1977) loop structures, and van Hoven (1982) has given an overview of the structure and stability of coronal loops using a force-free model of the magnetic field (Chiuderi *et al.*, 1980). We might remark at this point that although the concept of an isolated 'flare loop' has enjoyed much popularity over the years, there is no reason to believe that the magnetic field is absent outside of the radiating structures – we simply see certain magnetic field lines 'highlighted' by the presence of dense, radiating plasma within the bundle of field lines. Indeed, in order to prevent rapid expansion of the observed structures under the influence of the large magnetic pressure they contain [see equation (3.105)], external magnetic fields must be present.

Another major consequence of the frozen-in property is the extreme difficulty of fusing magnetic field lines together to release their energy (see Chapter 7). Translated into laboratory terms, the low resistivity of the solar plasma restricts the dissipation of electrical energy into other forms, such as thermal energy (Joule heating). This has proven to be a major obstacle to an entirely satisfactory theory of the solar flare process, using magnetic energy as a source.

6.2 Possible energy sources

It is generally accepted that the origin of a solar flare's energy is in the energy associated with stressed (i.e., nonpotential) magnetic fields. At this point, rather than attempt to prove this statement directly by investigating the structure of the solar magnetic field in flaring regions, it is perhaps equally persuasive to use arguments which show that there is simply no other conceivable energy reservoir of sufficient capacity. Thus, our heuristic proof will somewhat parallel the arguments used toward the turn of the century to show that nuclear fusion must be the source of stellar energy, i.e., by systematically dismissing all other possibilities (gravitational energy, thermal energy, chemical energy, etc.). Later in this chapter we will discuss evidence of a more direct nature that points to magnetic free energy as the driver of a solar flare.

Consider that a typical, moderately large solar flare extends over an area $S = 10^{18}$ cm^2 (as observed in chromospheric brightenings such as Hα) and over a depth corresponding to a columnar mass m of order 10^{-2} g cm^{-2} [e.g., Machado and Linsky (1975); Machado *et al.* (1980)]. The mean temperature of the particles in the volume in its pre-flare state is certainly less than the active-region coronal temperature, T_{cor}, viz. 3×10^6 K, and is probably in fact considerably less than this, since most of the pre-flare material is at chromospheric temperatures, T_{chrom} (10^4 K). A reasonable estimate of the thermal energy in the pre-flare region is therefore

$$E_{\text{ther}} = \frac{3S}{m_{\text{H}}}(m_{\text{cor}} k T_{\text{cor}} + (m - m_{\text{cor}}) k T_{\text{chrom}}),$$ (6.1)

where m_{cor} is the columnar mass of coronal material. Substituting numerical values $m_{\text{cor}} = 3 \times 10^{-6}$ g cm^{-2} (Vernazza *et al.*, 1976), $T_{\text{cor}} = 3 \times 10^6$ K, and $T_{\text{chrom}} = 10^4$ K, we find

$$E_{\text{ther}} = 3 \times 10^{28} \quad \text{erg},$$ (6.2)

which is insufficient to explain all but the smallest of flares. In any case, to provide the energy of a flare from pre-flare, thermal energy is a manifest violation of the Second Law of Thermodynamics – some form of nonthermal energy input is clearly needed. The estimate (6.2) is, however, useful in what follows.

Another candidate is gravitational potential energy, thus invoking some form of gravitational infall to provide the energy for the flare [see, e.g., Hyder (1967*a*,*b*)]. The energy available can be calculated from the expression

$$E_{\text{grav}} = Sg_0 \int h \, dm,$$ (6.3)

where h is the height of the flaring material and $g_0 = 2.74 \times 10^4$ cm s^{-2} is the acceleration due to gravity at the solar surface. Substituting numerical values, we find

$$E_{\text{grav}} = 3 \times 10^{28} \quad \text{erg},$$ (6.4)

which again is too small by several orders of magnitude. Furthermore, since much of the energy of a flare is contained in the coronal mass ejection (Chapter 9), involving an increase in the gravitational potential energy of the system, it is clear that gravitational energy cannot be responsible for a solar flare. We note parenthetically that the closeness of the estimates (6.2) and (6.4) for thermal and gravitational energy contents, respectively, of the pre-flare plasma is a direct consequence of the virial theorem for an atmosphere in hydrostatic equilibrium [see, e.g., Rose (1973)].

Having excluded the above candidate energy sources, we are now left with only nuclear and electromagnetic energy as possibilities. Temperatures in the outer part of the pre-flare solar atmosphere are much too low to support nuclear burning of any appreciable amount (note, however, that nuclear spallation can occur in the flare itself; see Section 5.7.3). We are thus forced to conclude that solar flares derive their energy from an electromagnetic reservoir. The relevant source of energy is in the magnetic field **B**, with an associated energy density $\mathbf{B}^2/8\pi$ erg cm^{-3}. Integrated over a flare volume of some $10^{27}\,\text{cm}^3$, this energy evaluates to $\approx 4 \times 10^{25}\mathbf{B}^2\,\text{erg}$, so that for $\mathbf{B} = 10^3\,\text{G}$, some $10^{32}\,\text{erg}$ of energy are realizable from the magnetic field. Although this is sufficient to account for the energy released in flares with observed volumes of this order, it remains to be shown that the energy can be released fast enough or stored over a long period of time without being released too soon. We return to this problem in Chapter 7.

6.3   Magnetic fields in active regions

As mentioned in Section 6.1, an active region can be defined as that part of the solar atmosphere where the strength of the magnetic field reaches values higher than in the surrounding, quiet regions. The strength of the field varies from one active region to the next, and it is not obvious that it is useful to give a lower limit for definition purposes. However, it may be worth noting that while the mean line-of-sight flux as observed with a low-resolution (e.g., 10 arc sec × 10 arc sec pixel) instrument is of the order of 1 G in the quiet Sun, it is an order of magnitude higher in active regions. We do not know the filling factor for magnetic fields in the solar plasma, but it is likely that it is quite small. Arguments have been advanced that the magnetic flux exists in thin discrete, subtelescopic ropes where the field strength may be of the order of 1500 G (Sheeley, 1966, 1967; Beckers and Schröter, 1968; Stenflo, 1973).

An independent, albeit indirect, method of assessing the magnetic field in an active region is to analyze the radio wave emission from the active-region plasma under the assumption that this emission is due to gyroresonance absorption at low harmonics of the gyrofrequency [e.g., Kundu and Lang (1985)]. Schmahl *et al.* (1982) examined the 6-cm radiation from an active region observed during the Solar Maximum Mission, and from simultaneous data in X-rays concluded that the optical depth for thermal Bremsstrahlung was too small to explain the radiation. On the other hand, gyroresonance absorption can lead to a sufficient optical depth at 6 cm. According to equation (5.69), if the radiation occurs at the third or fourth harmonic, the magnetic field in the 6-cm wavelength emitting plasma is 600 or 450 G, respectively.

In addition to knowing the strength of the magnetic field in an active region it is of the greatest importance to know the field configuration. A relaxed, potential field cannot provide any extractable energy, but must somehow be sheared or stressed. We have seen in Section 3.7 that magnetic fields in the low-gas-pressure environment will be governed by the force-free equation (3.106). Furthermore, observations furnish overwhelming evidence that the fields occur in the forms of arches or loops. We shall now look at a few examples of such fields.

The magnetic loop structures in the solar atmosphere are frequently observed to have a high-aspect ratio; i.e., the radius of the loop is much less than its lengths ('spaghetti strand'). Under such conditions we may neglect curvature effects and approximate the loop by a straight cylinder. Let the coordinate system describing such a field be (r, ϕ, z), with the Z-axis along the cylinder and (r, ϕ) being standard polar coordinates in the $z = $ constant planes. We shall consider axisymmetric, uniform fields, i.e., $\partial/\partial\phi = \partial/\partial z = 0$. From the condition $\nabla \cdot \mathbf{B} = 0$, it follows that $\partial(rB_r)/\partial r = 0$, and since rB_r vanishes on the axis, it must vanish everywhere, so that $B_r \equiv 0$. Then, evaluating $\nabla \times \mathbf{B}$, we find

$$\nabla \times \mathbf{B} = \left(0, -\frac{dB_z}{dr}, \frac{1}{r}\frac{d}{dr}(rB_\phi)\right), \tag{6.5}$$

and the r component of the force-free equation (3.106) becomes

$$B_z \frac{dB_z}{dr} + \frac{B_\phi}{r}\frac{d}{dr}(rB_\phi) = 0, \tag{6.6}$$

with all other components being identically zero. Equation (6.6) may be written

$$\frac{d}{dr}(B_z^2 + B_\phi^2) + \frac{2B_\phi^2}{r} = 0. \tag{6.7}$$

The solution of this equation is straightforward. Let

$$B_\phi^2 + B_z^2 = f(r), \tag{6.8}$$

where $f(r)$ is a generating function; then (6.7) gives

$$B_\phi^2 = -\frac{r}{2}\frac{d}{dr}[f(r)], \tag{6.9}$$

and, so, by (6.8),

$$B_z^2 = f(r) + \frac{r}{2}\frac{d}{dr}[f(r)]. \tag{6.10}$$

The function $f(r)$ is arbitrary, except for the requirement that both B_ϕ^2 and B_z^2 be positive semi-definite. This requires both $(d/dr)[f(r)] \leqslant 0$ and $2f + r(d/dr)[f(r)] \geqslant 0$;   i.e., $(2/r) + [1/f(r)](d/dr)[f(r)] \geqslant 0$;   i.e., $d(\ln f)/d(\ln r) \leqslant -2$. In other words, $f(r)$ must be a monotonically decreasing function which falls off with r no faster than $1/r^2$.

An example of such a function is

$$f(r) = \frac{B_0^2}{1 + (r/r_0)^2}, \tag{6.11}$$

which yields the field

$$B = B_0\left(0, \frac{r/r_0}{1 + (r/r_0)^2}, \frac{1}{1 + (r/r_0)^2}\right), \tag{6.12}$$

whose corresponding value of the force-free parameter α [equation (3.107)] is

$$\alpha(r) = \frac{\nabla \times \mathbf{B}}{\mathbf{B}} = \frac{2}{r_0(1 + (r/r_0)^2)}. \tag{6.13}$$

This field has the form of a nested set of helical flux surfaces, with the pitch of the helix B_ϕ/B_z increasing linearly with radius r. This arrangement gives a so-called 'constant twist' field; see Figure 6.2. We see that radially adjacent field lines are 'sheared' with respect to each other since the ratio B_ϕ/B_z increases with r. We shall see in Chapter 7 that this is a situation favorable to magnetic reconnection and energy release.

Another example worthy of note here is the constant-α (linear) cylindrical force-free field. From the fundamental force-free equation $\nabla \times \mathbf{B} = \alpha \mathbf{B}$

Fig. 6.2. Uniform-twist field. The force-free field $\mathbf{B} = (0, B_\phi, B_z)$ given by equation (6.12) has the property that $B_\phi/B_z = r/r_0$, i.e., that the tangent of the pitch angle ψ (to the Z-axis) increases linearly with radius. The field thus has the property that fields at large r are given correspondingly large pitch angles, in such a way as to keep all field lines which intersect a given radius at same level z intersecting the common radii at other levels. The angle between the relevant radii is a measure of the twist of the field (per unit length) and can be seen to be uniform in this case.

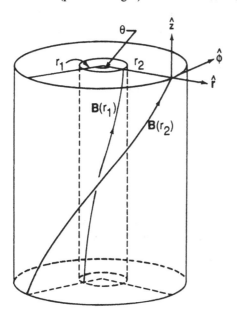

[equation (3.107)] and the expression (6.5) for $\nabla \times \mathbf{B}$, we find that this field must satisfy the relations

$$-\frac{dB_z}{dr} = \alpha B_\phi \tag{6.14}$$

and

$$\frac{1}{r}\frac{d}{dr}(rB_\phi) = \alpha B_z. \tag{6.15}$$

Elimination of B_ϕ gives

$$\frac{1}{r}\frac{d}{dr}\left(r\frac{dB_z}{dr}\right) + \alpha^2 B_z = 0. \tag{6.16}$$

This is a special case of Bessel's differential equation, with solution

$$B_z = J_0(\alpha r), \tag{6.17}$$

where J_0 is the Bessel function of order zero. By (6.14)

$$B_\phi = \frac{-1}{\alpha}\frac{dB_z}{dr} = -J_0'(\alpha r) = J_1(\alpha r), \tag{6.18}$$

where J_1 is the Bessel function of order one.

The solution (6.17) and (6.18) corresponds to a generating function

$$f(r) = J_0{}^2(\alpha r) + J_1{}^2(\alpha r), \tag{6.19}$$

as may be readily verified. It has the useful property that J_0 and J_1 are both alternating functions, so that reversals in field direction ('neutral surfaces') can occur. As will be seen in Chapter 7, such surfaces are suitable regions for rapid dissipation of magnetic energy.

It is difficult to avoid the conclusion that active regions are characterized by a host of strong magnetic fields, concentrated in flux tubes that form loops and arches of different sizes in the active-region atmosphere. This then is the environment in which flares occur, and, as we shall see in the following, it is precisely in these loops and arches that the flare energy release probably takes place.

6.4   Magnetic energy storage

If the flare energy comes from conversion of magnetic energy, the latter must have been stored somewhere and then suddenly released to produce the flare manifestations. Two possibilities for storage have been considered in the literature, viz. an *in situ* storage at the site of the flare and a storage of energy at a location distant from the flare site. To consider the more commonly discussed *in situ* storage, it is convenient to describe the process in terms of the magnetic field, $\mathbf{B}$. The remote storage scenario implies the concept of a global electrodynamic coupling of the solar plasma and is conveniently discussed in terms of the electric current density **j** (Alfvén, 1981).

6.4.1 *In situ storage*

When we treat the pre-flare conditions from a magnetohydrodynamic point of view, the electric current is eliminated from the equations and we treat the flare energy as stored in the magnetic field at the location where the flare will occur, i.e., in the corona above active regions. A large flare releases at least 10^{32} erg, and since we have seen that the only available source for this amount of energy in the corona is the magnetic field [cf. Parker (1957)], an *in situ* storage process is required that can accommodate such amounts of energy. However, the gas pressure in the corona surrounding the storage site is 2–3 orders of magnitude smaller than the magnetic pressure in the storage area, and special magnetic field configurations are required to build up the energy while maintaining stable conditions. A large number of investigations have been devoted to this fascinating problem, and most of them invoke the force-free magnetic field configuration; equation (3.106).

The storage of energy in this magnetic field can be thought of as a slow process where the field evolves through a sequence of force-free configurations, each time ending up in a higher energy state (Low, 1982). This slow process is driven by the motions in the low atmosphere (photosphere and lower chromosphere) where the energy of the plasma motions dominates the magnetic energy and where, therefore, the field is swept passively along with the plasma. This situation is characterized as a high-β plasma; i.e., the parameter

$$\beta = \frac{\text{gas pressure}}{\text{magnetic pressure}} = \frac{16\pi nkT}{B^2} \tag{6.20}$$

is large; $\beta \gg 1$. Higher up in the atmosphere (the corona) where the density is so small that the magnetic pressure dominates, we have a low-β plasma, and the magnetic field must take on a force-free character (Gold, 1964) as it slowly evolves. This MHD process is possible since the timescale for the 'wind-up' is days or weeks, while the field adjustment at any stage takes place with the Alfvén speed [equation (3.61)], leading to timescales of the order of seconds. This quasi-steady treatment of successive magnetostatic equilibria is certainly only a first approximation, but the full dynamic treatment is beyond present capabilities.

If $\alpha(\mathbf{r}, t)$ is constant in time and space, we can take the curl of equation (3.107) and find

$$\nabla^2 \mathbf{B} + \alpha^2 \mathbf{B} = 0, \tag{6.21}$$

which is the vector Helmholtz equation. This linear problem then is completely solvable (Chandrasekhar and Kendall, 1957; Ferraro and Plumpton, 1966; Nakagawa and Raadu, 1972).

Using equation (6.21) and observed boundary conditions in the form of the value of the longitudinal magnetic field (from magnetograph observations), we can compute the structure of the force-free field and the resulting stored magnetic energy. In these cases, α is adjusted until reasonable agreement is obtained with observations. With the availability of complete vector magnetic field observations, the field calculations have improved since α can be determined with higher precisions from equation (3.107) with the expression

$$\alpha = \frac{(\nabla \times \mathbf{B}) \cdot \mathbf{B}}{B^2}. \tag{6.22}$$

There is, of course, no real reason why α should be constant, either in space or in time. When $\alpha(\mathbf{r})$ is not constant, equation (3.107) is nonlinear, and the problem becomes very complex. No standard method of solutions exists, and the solutions one finds are not necessarily unique, rendering the physical interpretation difficult at best [see, e.g., Low (1982); Spicer (1982), and references therein].

6.4.2   Remote energy storage

Ionson (1982) applied the concept of global electrodynamic coupling – where an electric circuit transports energy from a plasma moving across a magnetic field to a remote site where the energy can be released – to the problem of heating the corona. Spicer (1982) and Hénoux (1983) have considered remote energy storage in the case of flare-energy release.

The flaring region is considered part of a magnetic loop [Figure 6.3(a)], and the physical characteristics are illustrated in Figure 6.3(b) by an electrodynamic circuit. The electrodynamic force is generated in the cooler layers of the photosphere where the degree of ionization is low enough so that sufficient neutral atoms are available to carry the positive ions across magnetic field lines. The electrons are more closely bound to the field lines and cannot remove the resulting imbalance of electric charge. This use of the dynamo model for flares (Heyvaerts, 1974) leads to an electromotive force

$$V_{\mathrm{emf}} = \int (\mathbf{v}_n \times \mathbf{B}) \, dx, \tag{6.23}$$

where v_n is the velocity of the neutral atoms. The dynamo, therefore, has the vast energy reservoir of the (sub) photospheric motions as its ultimate source.

To see the global aspect of this storage mechanism, we can use the equation for the total electric current, I, of the equivalent circuit in Figure 6.3(b), viz.

$$V_{emf} = RI + L\frac{dI}{dt}, \tag{6.24}$$

where R and L are the total resistance and self-inductance, both assumed independent of time. The solution of (6.24) gives

$$I = \frac{V_{emf}}{R}\left[1 - \exp\left(-\frac{Rt}{L}\right)\right]. \tag{6.25}$$

The electromagnetic energy stored in this circuit is

$$W_s = \tfrac{1}{2}LI^2 = \frac{1}{2}\frac{LV_{emf}^2}{R^2}\left[1 - \exp\left(-\frac{Rt}{L}\right)\right]^2, \tag{6.26}$$

and the timescale for energy storage is $\tau_s = L/R$. The resulting Ohmic heating, i.e., the integrated Joule heating [equation (3.47)] over the circuit, is

$$W_H = RI^2 = \frac{V_{emf}^2}{R}\left[1 - \exp\left(-\frac{Rt}{L}\right)\right]^2. \tag{6.27}$$

Hénoux (1983) used equations (6.23) and (6.26) to estimate the electromagnetic force and stored energy, assuming 10% dynamo efficiency,

Fig. 6.3. Remote energy storage – equivalent electrodynamic circuit: (a) electric current, I, flowing in magnetic loop between the photosphere and the corona; (b) equivalent L, R circuit.

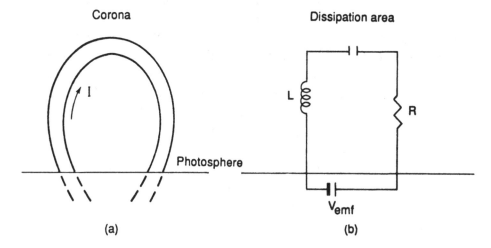

Corona   Dissipation area

Photosphere

L   R

V_{emf}

(a)   (b)

in a hypothetical loop situation with the following parameters: $v_n = 1 \, \text{km s}^{-1}$, $B = 100 \, \text{G}$, width of loop 1000 km, thickness of photospheric layer involved in the generator 1000 km, neutral particle density $n_n = 10^{18} \, \text{cm}^{-3}$, loop diameter 10 000 km, and resistivity perpendicular to magnetic field in the photosphere 0.1 Ohm m^{-1}. He found $V_{emf} = 10^7$ volts, $R = 10^{-5}$ Ohm and with $L = 1$ Henry, $\tau_s = 1$ day and $W_s = 10^{30}$ erg, while with $L = 10$ Henry, $\tau_s = 1$ week and $W_s = 10^{31}$ erg. These values are not inconsistent with values needed to support current views of the pre-flare timescale and stored energy associated with pre-flare conditions.

6.5   Precursors and flare buildup

Many flares are preceded, minutes to hours, by observable activity in or around the area where the flare will occur. We refer to such activity as *flare precursors*. The great interest in flare precursors – manifested in the form of special workshops on the topic as well as in numerous articles in recent years – is due both to the potential use of precursors in flare prediction and to the intriguing physical processes that may cause the precursor manifestation.

6.5.1   *Prominence activation*

One of the earliest documented flare precursors is the activation of prominences [e.g., Martin and Ramsey (1972)]; i.e., morphological and brightness changes in active-region prominences (active filaments) several to many minutes before the first Hα flare brightenings are observed. In some cases no emerging magnetic flux is seen that can be held responsible for the activation of the prominence. The activation therefore probably is to be considered the response to motions associated with changes in the adjacent, large-scale magnetic structure (Kundu *et al.*, 1985). In other cases it has been argued [e.g., Rust (1976)] that emerging magnetic flux may trigger destabilization of the magnetic field supporting the prominence, leading to its activation. This latter process is often observed during ejection of active-region filaments in the impulsive phase of flares; see Chapter 7.

In a thorough study of the well-observed flare of 21 May 1980, de Jager and Švestka (1985) gave an excellent account of the filament activation which preceded this large flare.

We conclude that the reason why a filament acts as a precursor to a flare seems to be the fact that the supporting magnetic field structure is ridding itself of stress and releasing energy. As this process proceeds, one should, therefore, at times be able to observe a change in the shear of the field, and

Fárník *et al.* (1983) did observe a decrease in magnetic shear during the activation of a filament some 15 min before a large two-ribbon flare.

6.5.2   *Pre-flare heating and brightening*

If the energy is not initially released in a sudden burst but leaks out as the magnetic field slowly starts to be rearranged, before the impulsive phase, one may expect a slow heating of the plasma and a more gradual brightening of the active-region plasma. This seems to be exactly what is happening in many instances; an example is shown in Figure 6.4 where elevated soft X-ray and UV levels are observed several minutes before the cataclysmic onset of the impulsive phase of a flare; [see also Poland *et al.* (1982), Hernandez *et al.* (1986), Klein *et al.* (1987), and Machado *et al.* (1988*a*)].

This onset of a flare, that at times can clearly be observed few to several minutes before the impulsive phase abruptly starts, is of great importance for a complete understanding of the flare phenomenon. It is during this onset period that the magnetic field in the active region becomes unstable and starts to readjust, possibly in response to new emerging magnetic flux. This restructuring – and possible annihilation – of the field then leads to prominence activation and heating of the surrounding plasma; see also MacNeice *et al.* (1985), who have discussed in detail the behavior of a well-observed flare from its onset marked by pre-flare brightening and filament disappearance.

6.5.3   *Magnetic shear and velocity fields*

A vector magnetograph uses observations of the polarization structure of a magnetically sensitive spectral line (Section 2.7) to record both the longitudinal field, $B_\parallel$, giving contours of strength, polarity, gradients, and the location of neutral lines, as well as the transverse field, $B_\perp$, which can reveal the presence of magnetic shear. In a potential, unstressed magnetic field the transverse component crosses the neutral line of the longitudinal field at right angles; see Figure 6.5(*a*). An observed rotation of the transverse field direction relative to the neutral line therefore constitutes a measure of the magnetic shear and of the energy stored in the magnetic field; see Figure 6.5(*b*).

Hagyard *et al.* (1984) studied the magnetic field in an active region and found that if they measured the degree of shear by the angle between the transverse component of the observed field and of a potential field ($\nabla \times \mathbf{B} = 0$), defined by the observed line-of-sight field, the shear had maximal values in two points that later became the starting points for flares.

Consequently, by measuring the shear one has an indication of the likelihood of whether a flare will occur, a condition now used in flare predictions; see also Makita *et al.* (1985).

The photospheric plasma, in which the magnetic shear is measured, is

Fig. 6.4. Light curves of the Ca XIX (3.2 Å), hard X-ray, O V (1371 Å) and Fe XXI (1354 Å) emissions, showing considerable preheating in the flare plasma several minutes prior to the onset of the main hard X-ray burst (from Cheng *et al.*, 1985).

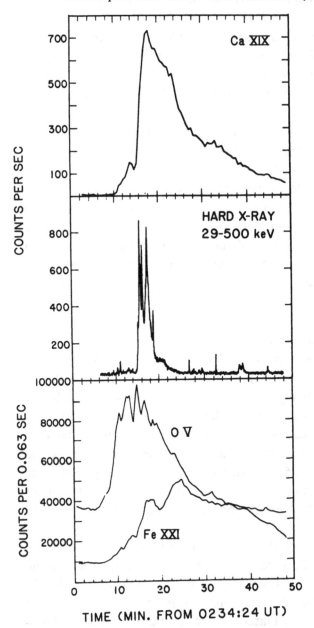

characterized by a large value of β [equation (6.20)]. This means that the magnetic shear can be created by appropriate velocity shears in the plasma, and Krall *et al.* (1982) found that pronounced velocity shears, coupled with alignment of the transverse magnetic field with the neutral line, accompanied major flares; see also Patty and Hagyard (1986).

Most flares occur in active regions with sunspots, and the more complex the sunspot distribution, the greater the incidence of flares (Giovanelli, 1939). The complexity refers both to the spatial distribution of magnetic polarity and to the temporal variations in magnetic intensity, as well as to the relative motions of the sunspots. These motions are likely to cause magnetic shear that subsequently leads to flare activity (Kovács and Dezsö, 1986). Magnetic flux appears and disappears as flux tubes buoy to the photosphere (Section 3.7) or sink down into the Sun. When this evolution is fast, flare activity increases (Bumba, 1986). We may expect that plasma motions and magnetic shears will be generated when magnetic flux emerges, and extra flare activity should result, as observed (Martin *et al.*, 1982). In addition, flares are known to be triggered when emerging flux interacts with pre-existing flux of the opposite polarity (Gaizauskas *et al.*, 1983; see also Priest *et al.*, 1986). This relationship is particularly strong for flares that

Fig. 6.5. Example of sheared photospheric magnetic fields in an active region. Solid (dashed) curves represent positive (negative) contours of the line-of-sight magnetic field. The directed line segments represent the strength (by length) and orientation (by direction) of the transverse component of the magnetic field. The field-of-view is 1.67 arc min × 1.67 arc min: (*a*) potential magnetic field calculated from the observed line-of-sight field; (*b*) observed magnetic field. Note the areas along the magnetic inversion line of the central bipole where the azimuth of the field deviates significantly from the potential field's azimuth. These areas were sites of flare kernels in the flares that erupted in this region (from Hagyard *et al.*, 1984).

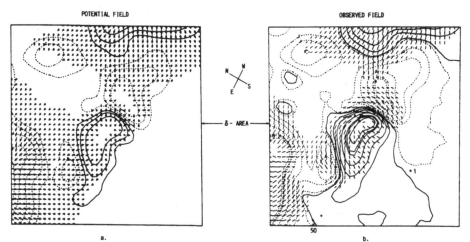

straddle the boundary between the new and the pre-existing flux regions (Martin *et al.*, 1984), probably illustrating the favorable conditions for magnetic reconnection in such geometry. Consequently, the most favorable conditions for flare occurrence should be realized when new flux erupts in the middle of an existing, complex spot group. Martres *et al.* (1968) seem to have been the first to recognize this situation, and termed the developing flux *structure magnétique évolutive*; see also Ribes (1969) and Martres *et al.* (1971).

There also seems to exist a relationship between flare occurrence and active-region prominences with a so-called 'pivot point'. When we observe quiescent prominences (filaments) for several months, on successive rotations of the Sun, the effect of the Sun's differential rotation is clearly seen: a more or less north–south oriented filament will eventually end up in a more east–west oriented position, because the parts of the filament closer to the equator rotate faster than the poleward parts. A north–south oriented filament at high northern latitudes will consequently, over several months, exhibit a counterclockwise rotation as well as an overall eastward drift relative to the (rigid) Carrington rotational velocity, as defined by central meridian rotation.

On the other hand, filaments close to active regions often do not exhibit differential rotation (Glackin, 1974). This anomaly has been studied further by Soru-Escaut *et al.* (1986) and Mouradian *et al.* (1988). They found that filaments on or near the magnetic neutral line in active regions show pivot points, limited areas which rotate rigidly, i.e., with the Carrington rotational velocity, independent of latitude. The rest of the filament, north, respectively south, of the pivot point, exhibits differential rotation, and consequently drift westward, respectively eastward, relative to the pivot point. It therefore looks as if the filament pivots on the rigidly rotating area – hence the name. Active centers normally form in and around pivot points, and these active centers have a higher flare productivity than other active centers (Soru-Escaut *et al.*, 1985). With the emergence of new magnetic flux the pivot point is destroyed and the filament may erupt (Martres *et al.*, 1986).

This sequence of events and the observation that pivot points occur on the neutral line lead us to assume that pivot points reveal areas that are magnetically connected to deep-seated structures in the magnetic field. The disappearance of the prominence is then due to a restructuring of the magnetic field, and if this restructuring also entails reconnection and annihilation, flare activity will result. We should consider the flare and associated prominence disappearance as two effects of a common, deeper-lying cause; see Section 7.1.7.

6.6 Bright points and microflares

We mentioned in Section 1.2 that flares come in many sizes, from the awesome importance 4 or class X flare to the smallest observable flares, the physics of which may, or may not, be of the same nature as their big counterparts. It has been known for some time that small short-lived brightenings in the solar atmosphere occur in both X-rays (Golub *et al.*, 1974, 1977) and the UV (Lites and Hansen, 1977; Emslie and Noyes, 1978; Bruner and Lites, 1979; Athay *et al.*, 1980; Brueckner, 1980).

These brightenings have been referred to as *bright points*, or microflares. We shall use both notations indistinguishably; the expression *miniflare* is also used. It is not obvious that we should always expect to see the emissions simultaneously in different wavelengths; e.g., all X-ray bright points may not also be observed as UV bright points. The very fact that a given bright point can be observed in one wavelength and not in another gives us a means to deduce information on parameters like temperature of the plasma involved. However, whether the bright point has a broad electromagnetic spectrum or not, we shall think of all these short-lived brightenings as belonging to the family of flares, making up the low-energy tail of the flare-distribution function [see Athay (1984)]. Details concerning the morphology and distribution of bright points have been treated by Sheeley and Golub (1979), Porter *et al.* (1984), and others.

Microflares in the transition-region plasma have been particularly well studied by Porter *et al.* (1984, 1986) observing the C IV 1548-Å line. These events occur on magnetic neutral lines throughout active regions and can be identified with small magnetic dipoles. Since some of these locations are later involved in a flare, while others are not, UV microflares are not a reliable bigger flare predictor. The UV events are impulsive, showing rise times typically of less than 20 s and sometimes as short as 5 s. The amplitudes range from the smallest detectable up to a factor of more than 50 seen in events classed as subflares.

Microflares in hard X-rays were discovered during a balloon flight by Lin *et al.* (1984). These microflares show peak fluxes 10–100 times less than normal flares. They are associated with small increases in soft X-rays and exhibit energy spectra that are well fitted by a power law ($E^{-3.0}$), indicating a nonthermal origin (see Chapter 7).

A scrutiny of the burst shape indicates that they may consist of small 'elementary' bursts, 100–1000 times smaller than the elementary bursts discussed by de Jager and de Jonge (1978). If this is correct, these elementary bursts may be related to the fine-scale microwave bursts observed by Kaufmann *et al.* (1985) at 15 GHz. All these observations indicate that energy release occurs nearly constantly in the active-region plasma on a very small scale. Subsequently, cataclysmic flare events may, or may not, take place.

Of greatest importance for our understanding of the physics of bright points is the information accumulated on the magnetic field pervading the bright point plasma (Withbroe *et al.*, 1985). It seems well established by magnetograph observations that a small emerging (from subphotospheric levels) magnetic dipole is formed where a bright point is seen in X-rays or in UV wavelength. The total magnetic flux involved is rather modest, probably 1000 times smaller than in an average active region, i.e., a few times 10^{19} Mx (Golub *et al.*, 1977), but annihilation (reconnection) may provide the required energy. It is therefore tempting to conclude that the energy necessary to heat the bright point plasma comes from the underlying magnetic field.

Sometimes only a small amount of magnetic energy is available – or, alternatively, only a small amount of available energy is converted to heat – and a microflare is all that results. At other times this first energy release in the form of a microflare may be followed by a more cataclysmic field annihilation, and a large flare occurs. The microflare is then to be considered a pre-flare phenomenon, a precursor (Section 6.5). An example of this behavior (Cheng *et al.*, 1984) is shown in Figure 6.6 where the upper panel shows the integrated Si IV 1402-Å line and continuum intensities before and during a flare, while the lower panel shows the variation of hard X-rays (29–57 keV) during the same period. The three Si IV, UV pre-flare bright points, labeled 1, 2, and 3 in the diagram, have no counterpart in the UV continuum or in hard X-rays. Such behavior is to be expected if the energy release takes place in the transition region where the Si IV line is formed, provided this release influences neither deeper atmospheric layers where the continuum emission originates nor those parts of flare loops responsible for the X-ray generation. On the other hand, we have numerous examples of bright points observable simultaneously in X-rays and Hα (Webb, 1985), a situation which shows that often the bright point structure contains both cool plasma (Hα) and energetic particles. Since the latter requires acceleration, we explain this situation by assuming that the magnetic field both heats the plasma and accelerates the required electrons. Schadee and Gaizauskas (1984) have presented convincing examples of somewhat larger Hα subflares that may also be observed as X-ray bright points. Since streams of accelerated electrons are responsible for Type III radio bursts [e.g., Kundu (1965)] one would also expect to observe at times 'radio bright points', i.e., a radio signature of microflares. There is evidence for a peak in the number of Type III bursts prior to large flares (Jackson, 1979; Kundu *et al.*, 1980), and Kane (1981) has analyzed Type III bursts that occur simultaneously with X-ray bursts (bright points).

Only a fairly small percentage (approximately 10%) of X-ray bright points are associated with Type III bursts, which indicates that only in those cases do the accelerated electrons have access to open magnetic lines of force out into the corona where they generate the Type III radio emission. In the other cases the electrons are trapped in the flaring bipolar magnetic loop

Fig. 6.6. Light curves of the integrated intensity of the Si IV 1402-Å line and the UV continuum (upper panel) and of the hard X-ray bursts (lower panel) during a flare of 12 October 1980. The three UV microflares seen in the Si IV line are indicated by the numerals 1, 2, and 3 (from Cheng *et al.*, 1984).

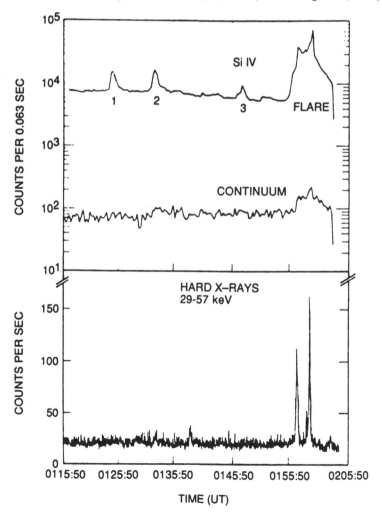

where they produce the Bremsstrahlung – mainly in the footpoints of the loop – recognized as X-ray bright points (Kundu *et al.*, 1980).

This completes our survey of the causes and manifestations of pre-flare energy buildup and release. We now turn to the explosive initial phase of the flare, where the energy, having been accumulated over many hours, or even days, is suddenly released.

7

The impulsive phase

This hitteth the nail on the head

Proverbs, J. Heywood (*c*. 1497–*c*. 1580)

The full-fledged 'canonical' flare is generally considered to be presaged by precursors, such as gradual soft X-ray brightenings (Section 6.5.2), starts with an impulsive phase, and develops into a more gradual phenomenon, classified as the main, thermal, or gradual phase of the flare, which finally decays. It is during the impulsive phase that the energy, which has slowly been built up and stored during the pre-flare period, is suddenly released and dissipated.

The impulsive phase is characterized by intense, rapidly fluctuating bursts of high-energy radiation, such as hard X-rays and γ-rays. There are also associated emissions at other energies, such as EUV and optical, generated as a result of the thermal response of the atmosphere to the rapid energization associated with this phase. Physically, the impulsive phase corresponds to the sudden release of stored magnetic energy into various forms, including accelerated particles, heating of plasma, bulk acceleration of fluid, and enhanced radiation fields. The degree to which each of the above energy forms is important, and the mechanisms for converting the magnetic energy into each of them constitute challenging questions of current research into not only solar flare physics, but other branches of astrophysics and laboratory plasma physics.

The fundamental size scale associated with magnetic energy release is of the order of a neutral sheet thickness (see Section 7.1.1). For solar parameters, this is orders of magnitude too small to be resolved observationally; hence, direct measurement of the energy-release process itself is impossible. Instead, solar physicists must probe the nature of this energy release by examining its signatures on observationally accessible scales and attempt to infer the nature of the energy release by synthesizing the available data into a coherent, self-consistent picture of the flare. In this chapter we shall examine these various observational manifestations of the impulsive phase and their interpretation in terms of canonical energy release and transport models. It is fair to say that at the present stage of our understanding, this interpretation has proceeded through comparison of the predicted behavior of certain 'paradigm' models (e.g., the thick-target model; see Section 7.2.2) with observations, resulting in constraints on the applicability and/or parameters of the model being established. A true discrimination between such paradigm models or, even better, a revision of a paradigm in response to observational requirements, is only now becoming within reach of both the theorist and the observer.

7.1   Energy release

As we have discussed in Chapter 6, the energy of a solar flare is stored in stressed (current-carrying) magnetic fields. In order to account for the impulsive energy release in the flare, we therefore require a mechanism that can rapidly dissipate the energy in these currents. The details of this mechanism and its geometry may also be determined by the requirement that suprathermal particles be accelerated (in order to produce impulsive phase hard X-ray bursts; see Section 7.2), or by the requirement that large-scale mass motions are established (see Chapter 9).

We noticed in Section 6.4 that the storage of magnetic energy in the solar atmosphere may be treated from a 'magnetic field picture' or from an 'electric current picture', depending on whether we analyze *in situ* or remote storage mechanisms. In a similar way the release of energy in the impulsive phase of a flare has been treated either in terms of magnetic field annihilation (magnetic merging) or in terms of electric field interruption (electric double layer).

7.1.1   *Magnetic reconnection*

The major problem faced by any mechanism which attempts to dissipate the stored magnetic energy is the low resistivity of pre-flare coronal plasma, with associated long magnetic diffusion times [equation (3.102)] and high magnetic Reynolds numbers [equation (3.99)], which requires the (frozen-in) field to move with the plasma and so inhibits its motion. Since the diffusion time scales as the square of the characteristic length scale l associated with changes in the magnetic field [equation (3.102)], it is clearly desirable to make this length l as small as possible. One way of doing this is to have two sets of oppositely directed fields in close proximity – a so-called neutral sheet configuration. Such a configuration could result, for example, near reversals in the constant-α force-free field discussed in Section 6.3 [see equations (6.17) and (6.18) and discussions following them]. Other topologies involve the appearance of one or more neutral points, where oppositely directed fields cross each other. In order to obtain a fairly clear picture of the physics of the dissipation of energy in such a configuration, through the process called magnetic reconnection, we shall first study the simplest case, that of a neutral sheet.

Consider the magnetic field geometry of Figure 7.1 where the magnetic field component B_x reverses direction along the line $y = 0$. For simplicity, we shall assume that the field geometry is independent of the coordinate z (out of the paper). If we take a closed path C as shown, we see that the line integral $\oint_C \mathbf{B} \cdot d\mathbf{l}$ has value $B_x \cdot 4L$, so that, applying Stokes' theorem,

$$\iint \mathbf{j} \cdot d\mathbf{S} = \frac{c}{4\pi} \iint (\nabla \times \mathbf{B}) \cdot d\mathbf{S} = \frac{c}{4\pi} \oint_C \mathbf{B} \cdot d\mathbf{l} = \frac{cB_x \cdot 4L}{4\pi},$$

and, since $\iint \mathbf{j} \cdot d\mathbf{S} \approx j \cdot (4Ll)$,

$$j \approx \frac{B_x c}{4\pi l}. \tag{7.1}$$

As l becomes smaller, j becomes as large as we please, so that even in the limit of very low resistivity η, the ohmic dissipation $\eta j^2 = (\eta B_x^2 c^2 / 16\pi^2)l^{-2}$ can become very large. For fields in vacuum, the limit on the smallness of l is the particle gyroradius [equation (4.20)] $r_B = mvc/qB_x$, where mv and q are the particle's momentum and charge respectively. In a plasma, however, the merging of oppositely directed field lines (and hence the decrease of l) is

Fig. 7.1. Neutral sheet magnetic reconnection. The magnetic field reverses direction along the X-axis, leading to a large current density in the Z-direction [equation (7.1)]. Inflow of the magnetic field at velocity v is balanced by fluid outflow at velocity v_x, since the fluid is incompressible. The dimensions of the current dissipation region are ($2L$) and ($2l$) parallel to the X- and Y-axes, respectively.

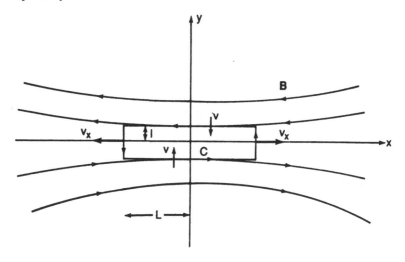

limited by the gas pressure p between the oppositely directed fields. Since the field and plasma are frozen together in the high conductivity limit [equation (3.98)], a reduction in l drives the plasma between the relevant pair of oppositely directed field lines to high densities. Such an adiabatic compression raises the gas pressure p between the fields and halts the compression when $p \approx B_x^2/8\pi$, i.e., when the gas and magnetic pressures are comparable.

The region of close, oppositely directed fields has only a finite length $2L$ (Figure 7.1). Outside of this region (i.e., at $|x| > L$), the gas pressure is substantially weaker than inside it ($|x| < L$), so that fluid is ejected along the field lines, reducing the builtup pressure and allowing the oppositely directed fields to approach closer, thereby increasing j and the ohmic dissipation rate. We can calculate, for the simple geometry of Figure 7.1, the reconnection velocity $v = dl/dt$. This will give us the rate at which field lines are swept into the reconnection volume, and hence the rate of energy dissipation in the process.

The momentum equation for the plasma which is ejected along the field lines (X-direction) is

$$\rho \frac{dv_x}{dt} = \frac{-\partial p}{\partial x},$$ (7.2)

which in a steady state ($\partial v_x/\partial t = 0$) becomes

$$\rho v_x \frac{\partial v_x}{\partial x} = \frac{-\partial p}{\partial x}.$$

Integrating from $x = 0$ to $x = L$ ($p = p_i$ to $p = p_o$), and considering the fluid to be incompressible [$\rho(x) = \text{const}$], we obtain

$$\tfrac{1}{2}\rho[v_x(L)]^2 = p_i - p_o,$$ (7.3)

where we have set $v_x(0) = 0$ by symmetry. The pressure difference $p_i - p_o$ is due to the magnetic field pressure $B_x^2/8\pi$ within the region $|x| \leqslant L$; thus

$$v_x(L) = \frac{B_x}{\sqrt{(4\pi\rho)}} = V_A,$$ (7.4)

which is simply the Alfvén speed [equation (3.61)] within the reconnecting region. Since the fluid is considered incompressible, continuity demands that the outward flow of material along the X-axis must be balanced by an inflow in the Y-direction, at the sought-after reconnection velocity v. Thus,

$$vL = V_A l$$ (7.5)

and

$$v = V_A \cdot \frac{l}{L}.$$ (7.6)

Having considered continuity and momentum balance, let us now close the solution by appealing to energetics. In steady state, the ohmic dissipation of energy in the reconnecting region must be just sufficient to balance the influx of magnetic energy $(B^2/8\pi)v$. We may therefore write

$$\iiint_V \eta j^2 \cdot dV = -\iint_S (B^2/8\pi)v \cdot dS, \tag{7.7}$$

where S is the surface bounding the cuboidal region V, formed by evoluting the curve C along the Z-direction. For the purpose of analysis, we can restrict the volume V to an arbitrary finite range of z. Thus, per unit length along the Z-axis, we have

$$\eta j^2 \cdot (4Ll) = \frac{B_x^2}{8\pi} \cdot v \cdot (4L),$$

or

$$\eta j^2 l = \frac{B_x^2 v}{8\pi}. \tag{7.8}$$

Substitution for j from Ampère's law (7.1) yields

$$v = \frac{\eta c^2}{2\pi l}. \tag{7.9}$$

Equations (7.6) and (7.9) can now be solved for the unknowns v and l, giving

$$v^2 = \frac{\eta c^2 V_A}{2\pi L} \quad \text{and} \quad l^2 = \frac{\eta c^2 L}{2\pi V_A}. \tag{7.10}$$

These results may be conveniently expressed in terms of a 'longitudinal' magnetic Reynolds number R_m for the reconnection region [equation (3.99)]

$$R_m = \frac{|\nabla \times (\mathbf{v} \times \mathbf{B})|}{\left|\dfrac{\eta c^2}{4\pi} \nabla^2 B\right|} \approx \frac{V_A B_x/(2L)}{\dfrac{\eta c^2}{4\pi} \cdot \dfrac{B_x}{(2L)^2}} = \frac{8\pi L V_A}{\eta c^2},$$

giving

$$v = 2V_A/R_m^{1/2} \quad \text{and} \quad l = 2L/R_m^{1/2}. \tag{7.11}$$

For comparison, we note that for fields which do not exhibit a neutral sheet such as in Figure 7.1, a scale length of L would be appropriate everywhere, and the reconnection velocity would be given by

$$v = L/\tau_D = L \cdot \left(\frac{\eta c^2}{4\pi L^2}\right) = 2V_A/R_m, \tag{7.12}$$

where τ_D is the magnetic diffusion time scale [equation (3.102)]. Thus, the presence of a neutral sheet enhances the reconnection velocity by a factor $R_m^{1/2}$. Values of R_m are quite large in the solar atmosphere, as noted in

Chapter 3; thus, the reconnection rate is considerably enhanced. However, the reconnection velocity is still significantly lower, by the same $R_m^{1/2}$ factor, than the Alfvén speed.

In terms of the Alfvén crossing time for the reconnection region, $\tau_A = l/V_A$, and the resistive diffusion time across the region, $\tau_D = 4\pi l^2/\eta c^2$, the reconnection time $\tau_R = l/v$ is given by

$$\tau_R = \frac{\tau_A}{2} R_m^{1/2} = \frac{\tau_A}{2}\left(\frac{2L}{l}\right)^{1/2} R_{m\perp}^{1/2} = \frac{\tau_A}{2} R_m^{1/4}(\tau_D/\tau_A)^{1/2},$$

where we have used equation (7.11). $R_{m\perp}$ is the 'transverse' magnetic Reynolds number given by equation (3.102) $R_{m\perp} = 4\pi l V_A/\eta c^2 = \tau_D/\tau_A$. The expression for τ_R simplifies to

$$\tau_R = \tfrac{1}{2} R_m^{1/4}\sqrt{(\tau_A\tau_D)}. \tag{7.13}$$

Thus, the reconnection time is the geometric mean of the Alfvén transit time and resistive diffusion time, multiplied by the factor $\tfrac{1}{2}R_m^{1/4}$ which is still considerable for solar conditions. Equation (7.13) is useful for comparison with later results in this chapter.

7.1.2   *Fluid stagnation and the Petschek model*

Petschek (1964) pointed out that since material is ejected symmetrically along the X-axis, the origin must be a fluid stagnation point ($v = 0$, $\nabla \cdot v > 0$). If we consider a parcel of incompressible, highly conducting fluid (with its associated frozen-in magnetic field B_{x0}) bounded by $0 < x < x_0, 0 < y < y_0$, as it is convected toward the neutral line (X-axis), then conservation of mass and magnetic flux (in the X-direction) demand that, at a time when the fluid has been compressed to a vertical extent y_1 (so that it now occupies the region $0 < y < y_1$), its horizontal extent is given by

$$\Delta x = \frac{x_0 y_0}{y_1}$$

and the magnetic field within the parcel is

$$B_x = B_{x0}\frac{y_0}{y_1}.$$

The velocity of the right-hand edge of the parcel is

$$v_x = \frac{d(\Delta x)}{dt} = \frac{x_0 y_0}{y_1^2} v,$$

where v is the reconnection velocity $= -dy_1/dt$.

We see from this analysis that as $y_1 \to 0$, both v_x and B_x increase without bound. In practice, this cannot occur, since the driving force is the external magnetic pressure $B_0{}^2/8\pi$, and therefore only values of B_x which are less than B_0 are allowed. In order to reconcile this with the above analysis, we require that the fluid velocity change from a compressive motion in the Y-direction to an expansion in the X-direction at some significantly positive value of y_1, with a corresponding dissipation of magnetic energy. This abrupt transition of fluid velocity gives rise to a standing shock profile such as is shown in Figure 7.2. In such a picture the magnetic field lines truly 'reconnect' with hitherto distinct field lines, rather than simply diffusing away – this is true 'magnetic reconnection'. The dynamics of the shock formation [e.g., Parker (1979)] implies that the horizontal extent, L, of the reconnecting region is reduced to the same order as l, the thickness of the reconnecting region. This significantly reduces the applicable Reynolds number in the reconnecting region and permits much faster reconnection, although admittedly only over a smaller volume. According to Petschek, the reconnection velocity can be as high as $V_A/\ln(R_m)$, where R_m is the magnetic Reynolds number for the large-scale field. .

The geometry of Figure 7.2 shows us that magnetic reconnection is more than simply ohmic heating due to induced currents j; rather, the actual topology of the magnetic field configuration is altered, in this case changing from 'concave up–down' to 'concave right–left'. The tension in these newly

Fig. 7.2. Petschek reconnection geometry. The solid lines and arrows represent magnetic field lines and the dashed lines and open arrows velocity streamlines. Note the discontinuity in velocity at the standing shocks S emanating from the reconnection region (innermost rectangle) and the reconnection of hitherto distinct field lines onto each other.

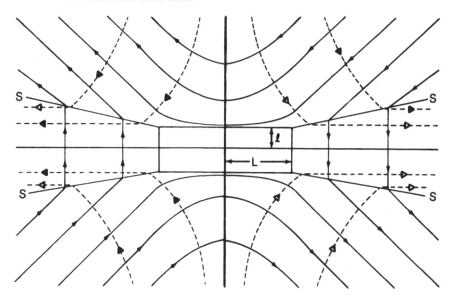

created field lines serves to propel the field lines and their associated frozen-in fluid outward from the reconnecting region at great speeds. This not only evacuates the reconnection region, permitting further rapid reconnection, but such rapid outward moving field lines may play a role in particle acceleration (Section 7.1.7). Recently, Priest and Forbes (1986) reviewed the various topologies of reconnection regions, including that of Petschek and that of Sonnerup (1970), who used two standing waves in each quadrant [see also Parker (1979)]. Priest and Forbes pointed out that all of the various modes of reconnection considered to date can be generalized into a comprehensive picture, which also admits other topologies not previously considered. We refer the interested reader to their article for details.

7.1.3   *Resistive instabilities and the tearing mode*

If a magnetic field configuration such as that of Figure 7.1 is subjected to a 'rippling' disturbance in the X-direction, reconnection may occur at several points along the neutral line. This leads to the formation of 'magnetic islands' between the reconnection points, as shown in Figure 7.3. Such magnetic islands, being isolated from the background field, tend to collapse under the action of field-line tension into smaller islands, thereby drawing material away from the reconnection sites at a substantial rate and enhancing the rate of dissipation of magnetic energy. For a rigorous treatment of this mode we draw the reader's attention to the original work by Furth *et al.* (1963) and to the application to solar flares by Spicer (1976, 1977). The essential result is that the reconnection time τ_R is given by

$$\tau_R \approx \tau_A^{2/5}\tau_D^{3/5}, \tag{7.14}$$

Fig. 7.3. Formation of magnetic islands between X-type neutral points (dots), as a result of a rippling disturbance in a neutral sheet geometry (cf. Figure 7.1). The tension in the closed field lines acts to increase the outflow velocity at the neutral points (cf. v_x in Figure 7.1) and so enhances the rate of magnetic energy dissipation.

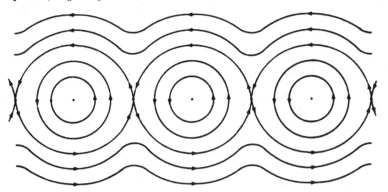

which is a substantial improvement over the rate given by equation (7.13) because of the absence of the $R_{\mathrm{m}}^{1/4}$ factor.

A significant point made by Spicer (1976, 1977) is that reconnecting regions may overlap as shown in Figure 7.4. This results in a large increase in the number of X-type magnetic neutral points, a corresponding increase in the volume of reconnecting regions, and a very rapid energy dissipation rate.

7.1.4 *Energy dissipation rates*

Let us now calculate typical energy dissipation rates for the scenarios considered so far to see if any of them are capable of releasing enough energy to account for the impulsive phase of a solar flare. The rate of energy release is given by

$$\frac{d\varepsilon}{dt} = \left(\frac{B^2}{8\pi}\right) \cdot V\tau_{\mathrm{R}}^{-1}, \tag{7.15}$$

where B is the pre-flare magnetic field strength, V the volume of the reconnecting region(s), and τ_{R} the reconnection timescale. We consider a pre-flare loop with length $L = 10^9$ cm, magnetic field strength $B = 300$ G, density $n = 10^{10}$ cm^{-3}, and temperature $T = 2 \times 10^6$ K. Note that as the flare proceeds, the increase in temperature will drive the resistivity down [by equation (4.82)] and cause the calculations to overestimate the actual energy dissipation rate, unless anomalously high resistivities (Section 7.1.5) are invoked. The opposite scenario – in which the increased densities associated with the compressed plasma in the reconnection region cause enhanced radiation, reduced temperature, and therefore an *increased* reconnection rate – has been proposed by van Hoven *et al.* (1983, 1984). Such a reduced temperature plasma is consistent with the ubiquitous presence of prominences around the time of flare-energy release (Section

Fig. 7.4. Nonlinear overlap of magnetic reconnection regions. Note the greatly increased number of magnetic neutral points (dots) (after Spicer, 1976).

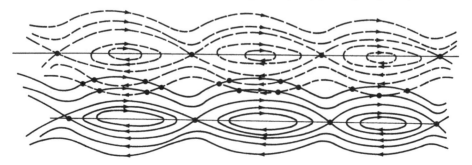

7.1.6). These considerations of time-varying resistivity do not, however, significantly affect our rough estimates below. For the parameters chosen, the Alfvén velocity $V_A = 6 \times 10^8 \, \text{cm s}^{-1}$ and the classical resistivity $\eta = 1.5 \times 10^{-7} \, T^{-3/2} = 5 \times 10^{-17}$ s. The magnetic Reynolds number R_m for the large-scale field is

$$R_m = \frac{4\pi L^2}{\eta c^2} \left(\frac{L}{V_A} \right)^{-1} = 10^{14}, \tag{7.16}$$

and, by equation (7.11), $l \approx 200$ cm. The timescales τ_A and τ_D are given by

$$\tau_A = \frac{l}{V_A} \approx 3 \times 10^{-7} \, \text{s} \quad \text{and} \quad \tau_D = \frac{4\pi l^2}{\eta c^2} \approx 10 \, \text{s}. \tag{7.17}$$

With these numerical values we may calculate $d\varepsilon/dt$ for each of the reconnection scenarios discussed above. First, for the simple, neutral sheet reconnection, for which $V \approx L^2 l \approx 2 \times 10^{20} \, \text{cm}^3$ and for which, by equation (7.13), $\tau_R = \frac{1}{2} R_m^{1/4} \tau_A^{1/2} \tau_D^{1/2} \approx 3$ s, we obtain

$$\left. \frac{d\varepsilon}{dt} \right|_{\text{neutral sheet}} \approx 2 \times 10^{23} \quad \text{erg s}^{-1}. \tag{7.18}$$

Second, for the single-region Petschek scenario, $V \approx l^3 \approx 10^7 \, \text{cm}^3$, $\tau_R \approx \tau_A^{1/2} \approx 5 \times 10^{-4}$ s, and

$$\left. \frac{d\varepsilon}{dt} \right|_{\text{Petschek}} \approx 2 \times 10^{13} \quad \text{erg s}^{-1}. \tag{7.19}$$

In the case of multiple reconnecting regions along the neutral line, such as in the tearing mode, the source volume increases again to $V \approx L^2 l \approx 2 \times 10^{20} \, \text{cm}^3$, while the reconnection time is, by equation (7.14), $\tau_R \approx \tau_A^{2/5} \tau_D^{3/5} \approx 10^{-2}$ s, and

$$\left. \frac{d\varepsilon}{dt} \right|_{\text{tearing mode}} \approx 7 \times 10^{25} \quad \text{erg s}^{-1}. \tag{7.20}$$

Finally, for the multiply overlapping tearing mode geometry of Figure 7.4, $V \approx L^2 a$, where a is the thickness of the flaring loop $\approx 10^8$ cm; i.e., $V \approx 10^{26} \, \text{cm}^3$ and

$$\left. \frac{d\varepsilon}{dt} \right|_{\text{multiple tearing}} \approx 3 \times 10^{31} \quad \text{erg s}^{-1}. \tag{7.21}$$

Apparently only the last of these is capable of accounting for the energy released during the impulsive phase of a large solar flare. However, we have, without justification, used the value given by classical resistivity due to Coulomb collisions in the development of the timescales τ_D and τ_R. Let us now investigate whether mechanisms to anomalously enhance η over this value can be present.

7.1.5 *Plasma instabilities and anomalous resistivity*

The fundamental problem which leads to the embarrassingly low rates of energy release [see equations (7.18)–(7.20)] is the low value of the classical resistivity in a solar plasma, which renders the frozen-in field approximation extremely good, and which limits the energy-release rate by effectively preventing field lines from diffusing through the surrounding plasma and subsequently reconnecting. An obvious way of enhancing the energy dissipation rates $d\varepsilon/dt$ is therefore to increase η over its classical value. One way of doing this is to introduce a level of plasma waves in the reconnection region. The presence of such coherent plasma oscillations has the effect of correlating particles over length scales much larger than the usual Debye length (Section 4.1), so enhancing the energy loss felt by a test particle colliding with one of these oscillating particles. (The test particle 'feels' the effect of all the other oscillating particles at the same time.) This can greatly enhance the rate of energy loss and so increase the collision frequency from its classical value v_c to a much enhanced value v_{eff}. Since $\eta \propto v_c$, we see that this process will accomplish the desired increase in the magnetic diffusion rate. It is well established that astrophysical plasmas have a wide variety of instability modes open to them [see, e.g., Melrose (1980)]; let us here, for simplicity, focus on one of these – the ion-acoustic instability (Section 4.5).

An ion-acoustic instability arises when the current density is raised to such a level that the associated drift velocity v_d exceeds the ion-acoustic velocity $c_s = (kT_e/m_p)^{1/2}$; see equations (4.113) and (4.114) with $T_e \gg T_i$. The combined electron-ion, two-fluid distribution function then exhibits a two-stream profile, leading to the generation of ion-acoustic waves and an enhanced v_{eff}. The current density j in a reconnection region is given by

$$j = n_e e v_d = \frac{c}{4\pi} (\nabla \times \mathbf{B}) \approx \frac{cB}{4\pi l}.$$

Ion-acoustic instability therefore occurs when

$$\frac{cB}{4\pi n_e l e} > \left(\frac{kT_e}{m_p}\right)^{1/2}, \tag{7.22}$$

which reduces to

$$\frac{B^2 r_B}{4\pi n_e l m_p} > \left(\frac{kT_e}{m_p}\right), \tag{7.23}$$

where

$$r_B = \frac{m_p c}{eB} \left(\frac{kT_e}{m_p}\right)^{1/2}$$

is the gyroradius for a thermal ion [cf. equation (4.20)]. The inequality (7.23) further reduces to

$$\frac{4r_{B}}{l} > \frac{2n_{e}kT_{e}}{(B^2/8\pi)} = \beta,$$

where β is the ratio of gas to magnetic pressures [equation (6.20)]. Thus, if

$$l < \left(\frac{4}{\beta}\right)r_{B}, \tag{7.24}$$

the plasma resistivity η will become anomalously enhanced, and the energy dissipation rate will be correspondingly increased. For the pre-flare loop parameters discussed earlier $r_{B} \approx 4.5$ cm and $\beta \approx 1.5 \times 10^{-3}$; thus, the condition (7.24) becomes

$$l \lesssim 10^4 \text{ cm.}$$

This condition is apparently satisfied by the current sheet parameters discussed above, and we may expect the generation of unstable currents with an associated enhancement in the resistivity η. However, we recall from equation (7.10) that as η increases, so does the thickness of the reconnecting region, reducing $\nabla \times \mathbf{B}$ and the current density $\mathbf{j}$. What we therefore expect is for the system to evolve to a state of marginal stability where the thickness of the region is such that the currents produced in the reconnection region set up a resistivity consistent with its thickness. We refer the interested reader to the work of Spicer (1976, 1977) for details on the relevant physics; the end result is an enhancement in energy release of some 1–2 orders of magnitude over the case of classical resistivity.

In summary, therefore, magnetic reconnection in a multiply overlapping nonlinear tearing mode configuration appears to be a favorable mechanism to account for the rapid release of magnetic energy in the impulsive phase of a solar flare. The presence of anomalous resistivity may substantially enhance the energy release rate.

7.1.6   *Topology of the energy-release region and flare models*

In discussing the energy release in this chapter, we have identified the basic physical process as magnetic reconnection through, for example, a tearing mode instability. However, we do not know the details of how the appropriate conditions are set up for reconnection to occur. This subject is usually referred to by the title of 'flare models'. An excellent review of such models has been given by Sturrock (1980), and we refer the interested reader there for a more thorough discussion. Here we summarize the salient features of several flare models which have been offered over the years.

The magnetic topology of the reconnecting region is what basically defines the flare model, and this may not be the simple neutral sheet configuration discussed above. Indeed a variety of reconnection models have been proposed over the years. Sweet (1958) envisioned the merging of two bipolar flux regions with the formation of a neutral sheet in between (see Figure 7.5). This model was also presumed by Syrovatskii (1966, 1969). Gold and Hoyle (1960) modeled the merging of two magnetic flux tubes in an orthogonal direction, as depicted in Figure 7.6. We note from the figure that the longitudinal (toroidal) currents are parallel, causing the loops to mutually attract. The transverse (poloidal) magnetic fields in the region of merging are, however, oppositely directed, and this gives rise to the neutral sheet configuration of Figure 7.1. Sturrock (1968) suggested the 'helmet streamer' configuration in Figure 7.7 with a neutral sheet formed above the Y-type neutral point. Since streamer-like structures are frequently observed by white-light coronagraphs, Sturrock's model appears to be a viable scenario, particularly in events involving the ejection of material into interplanetary space (Chapter 9) along open field lines above the Y-type neutral point.

Fig. 7.5. Merging of two bipolar regions D_1 and D_2 to produce a neutral current sheet S (after Sweet, 1958).

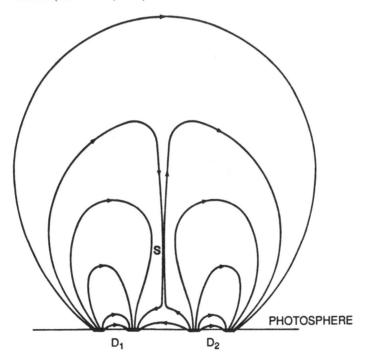

Fig. 7.6. Gold–Hoyle model of a solar flare. Two flux tubes carry similar twists (currents) and so attract each other. The longitudinal fields are antiparallel, and therefore so are the poloidal fields at the boundary. This creates a neutral sheet configuration at the surface where the two tubes merge.

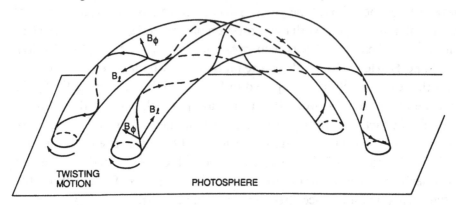

Fig. 7.7. Solar flare geometry envisaged by Sturrock (1968). The central helmet streamer configuration (see Figure 9.1) creates a neutral sheet terminating in a *Y*-type neutral point, which is the region of initial reconnection. The field lines below the *Y*-point are closed, and produce the hard X-ray and Hα emissions, etc.; the field lines above the *Y*-point are open and are responsible for particle ejecta.

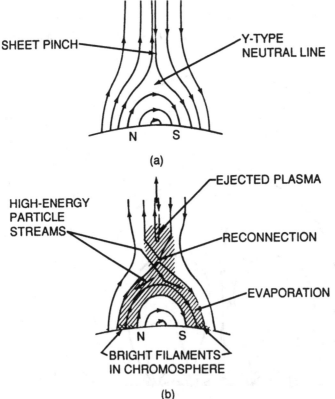

Heyvaerts *et al.* (1977) suggested that flares result when new flux emerges from below the photosphere by virtue of its magnetic buoyancy (see Section 3.7) and collides with an existing magnetic flux tube creating a neutral sheet at the common boundary of the flux tubes (see Figure 7.8). The characteristic quadrupole field configuration associated with such a model has been observed in some events [e.g., Velusamy and Kundu (1982)], and since emerging flux is often seen associated with the occurrence of flares, the model of Heyvaerts *et al.* seems pertinent in many cases. Finally, Spicer (1976, 1977) suggested that magnetic kink instabilities in a toroidal arch may produce overlap of magnetic field lines such as in the tearing mode formation discussed above. This model has the virtue of explaining compact flares without significant restructuring of the surface magnetic field.

7.1.7   *The flare-trigger mechanism*

Many flares have filament eruptions associated with them, as evidenced by Hα movies which show the ejection of cool absorbing material outward from the flare site (see Chapter 9). Kopp and Pneuman (1976) suggested that the eruption of this filament (due to a magnetic instability, perhaps) would 'draw out' field lines, as shown in Figure 7.9, leading to the required neutral sheet magnetic field configuration.

This scenario implies a more global nature of the flare-triggering mechanism. The observed mass ejections from certain flares (see Section 1.3) may not simply be considered the result of an expelling force due to the flare; the mass ejections often seem to start simultaneously with, or even before, the main flare energy release, and large areas of the surrounding corona are

Fig. 7.8. Emerging flux model. A new flux tube rises and collides with an existing flux tube, creating a current sheet. During the impulsive phase the reconnection is a 'driven' process, but relaxes to a steady state during the main phase.

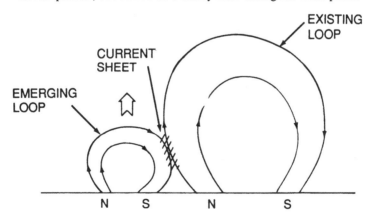

Fig. 7.9. Model of Kopp and Pneuman (1976). A filament eruption (*a*) (see Chapter 9) 'tears open' field lines producing an open sheet-type configuration (*b*). The reconnection process causes field lines to reconnect at progressively higher points (*c*). This produces a chromospheric signature in which the bright emission moves away from the neutral line – the classic 'two-ribbon' flare picture.

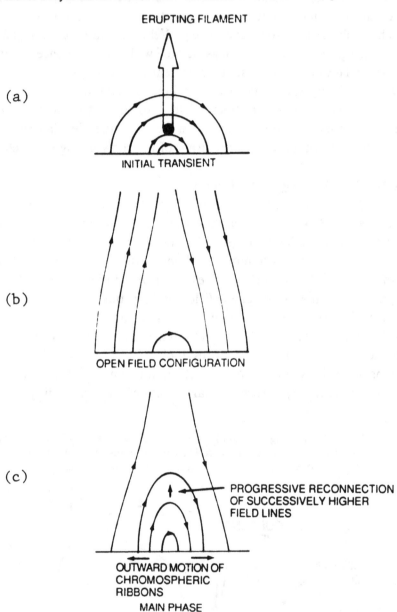

immediately involved. Similarly, the destabilization of an active-region prominence, or of a quiescent prominence, is generated by a triggered realignment of the whole magnetic structure (Section 6.5.3). This realignment may at times be violent enough to lead to an eruption of the magnetic field (Rompolt, 1984). We conclude that it is the magnetic structure, both at the site of the energy release and in the surrounding plasma, that sets up the conditions for releasing energy (by reconnection) and ejecting mass. If the global field is closed above the flare site, the material is confined, and we observe a compact, confined flare. If the field is open, matter may be expelled, and a coronal mass ejection takes place. Machado *et al.* (1988*a*) referred to this class as *ejective* flares. According to these authors the basic magnetic field structure necessary for a flare to occur is a so-called initiating closed bipole plus one or more adjacent closed bipoles. The impulsive phase then starts when the initiating bipole strongly interacts with the adjacent bipole or bipoles. However, the energy release need not occur solely in the neutral sheet; the energy in the stressed bipole fields can also be released. In this picture a flaring bipole is confined or ejective, depending on the global configuration of the magnetic field in the corona around the various bipoles in the flare site.

7.1.8 *Particle acceleration*

In order to produce hard X-ray and γ-ray bursts, the basic flare mechanism must be capable of accelerating copious quantities of particles, both electrons and protons. Further evidence for such particle acceleration can be adduced from interpretation of the characteristics of the secondary response of the atmosphere to the energy release (Chapter 8) and from the appearance of microwave bursts (Section 7.3), Type III radio bursts (Section 7.6.2), and interplanetary particle streams (Section 7.6.3). The problem of particle acceleration in flares is therefore a fundamental one, yet it is one that has so far eluded a completely satisfactory explanation. Indeed, one of the motivations for study of the radiation signatures of the impulsive phase and their properties is to constrain the parameters (number, spectrum, etc.) of the particle acceleration process. Furthermore, we have every reason to believe that acceleration processes similar to those in solar flares operate in a wide variety of other astrophysical plasmas, from planetary magnetospheres to active galactic nuclei, and this is another reason why solar flares are so intensely studied.

In this section we shall address three of the main mechanisms proposed for particle acceleration in the impulsive phase of solar flares – double layers, shocks, and direct electric field acceleration.

7.1.8.1   *Double layers*

Alfvén and Carlqvist (1967) proposed a flare model in which an ever-increasing current is forced longitudinally along a flare loop, for example, by constantly increasing the stress of the magnetic field. As the current reaches a critical value, plasma instabilities, such as the ion-acoustic instability, develop and suddenly enhance the resistivity, forming a large potential drop in this region. As a result, a so-called *double layer* forms and is responsible for the runaway acceleration (Section 7.1.8.3) of the high-velocity portion of the particle population. Double layers may be important in many areas of astrophysics (Williams, 1986; Alfvén, 1987) and to understand the physics involved one may use a particle model (Lyons and Williams, 1985) or a circuit theory model (Alfvén, 1981).

A double layer (see Borovsky, 1983) is a self-sustaining system. Suppose the electrostatic potential ϕ drops from left to right, over a distance of a few Debye lengths, from a value corresponding to several times the thermal energy of the plasma to a much smaller value (without loss of generality, this may be taken as the zero reference point; see Figure 7.10). Protons arriving from the left are accelerated by the potential drop and electrons are reflected by it. As the protons pick up speed, continuity demands that their number density n_p decrease in order that the flux $n_p v$ be conserved across the potential drop. Thus, the number density of such protons decreases from left to right. To these must be added the number density of protons arriving from the right and reflected by the large *increase* in potential that they see. On the other hand, electrons arriving from the right are attracted by the positive potential and are accelerated; thus, their number density n_e decreases from right to left across the potential difference (see Figure 7.10).

The net result of these varying proton and electron densities is for the proton density to be higher in the left half of the potential drop than in the right, and vice versa for the electrons; see Figure 7.10. The resulting charge density is responsible, through Poisson's equation $d^2\phi/dx^2 = -4\pi e(n_i - n_e)$, for sustaining the shape of the potential drop, and the curve $\phi(x)$ is concave downward in the left half of the drop, where $n_i > n_e$, and concave upward in the right half, where $n_i < n_e$.

Typical values of $(n_i - n_e)$ can be up to half the ambient plasma density, and the potential drop ϕ is set by the length Δx by which the charge layers are separated. The situation closely resembles a parallel-plate capacitor with equal and opposite charges on plates separated by the distance Δx; hence, the name 'double layer'. Since the voltage drop across a capacitor with fixed charges on the plates increases as the plate separation Δx increases (the electric field between the plates being approximately constant), we see that

Δx cannot be too large; otherwise, the huge potential difference would bring the charge layers closer together by electrostatic attraction. Thus, the potential drop $\Delta\phi$ must occur over a few Debye (shielding) lengths. Double layers are thus extremely localized regions wherein highly efficient particle acceleration can occur. Once established, they are self-sustaining; however, the problem in solar flare physics is one of creating them in the first place.

7.1.8.2 *Shock acceleration*

The idea of shock acceleration was first proposed as a mechanism to explain the acceleration of very high-energy protons in order to account for

Fig. 7.10. Schematic of a double layer. As a result of the large potential drop, particles are accelerated by tens of kT. Ions are accelerated to the right, electrons to the left. By continuity, the high velocity streams have a reduced density, so that the difference in electron and ion densities creates a charge density which sustains the potential drop.

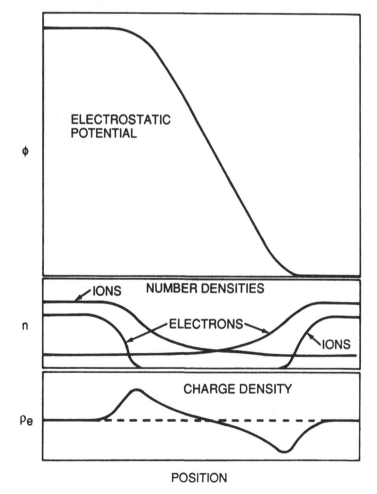

high-energy γ-ray bursts (see Section 7.4). Originally, shock acceleration was proposed as a 'second-stage' acceleration process, whereby the redistribution of the energy released during the impulsive phase creates the shocked environment (Section 7.1.2), which can then act on particles preaccelerated to moderate energies in the impulsive phase and accelerate them to the very high energies required (several hundred kiloelectron volts for the electrons and tens of million electron volts for the protons and ions). Observations from SMM revealing the synchronism of hard X-ray and γ-ray bursts (see Section 7.4) have now shown this to be wrong: shock acceleration is an intrinsically impulsive phase phenomenon.

As an example of shock acceleration, let us discuss the first-order Fermi acceleration process which has been proposed by Bai *et al.* (1983) as a mechanism for relatively prompt acceleration of protons in solar flares. Consider a particle trapped in a region between two converging shock fronts [which may be considered as having originated in two reconnecting regions (Section 7.1.2) or at the footpoints of the flare loop as a result of impulsive energy deposition by nonthermal electrons; see Section 8.1.1]. A particle moving to the right at velocity v is reflected by the leftward-moving shock, which has a velocity $(-u)$ (Figure 7.11). In the frame of the shock, the particle's velocities before and after reflection are $(v + u)$ and $-(v + u)$, respectively; in the laboratory frame, they are v and $-(2v + u)$. The gain in the energy of the particle as a result of the reflection is

$$\Delta E = \tfrac{1}{2}m(2v + u)^2 - \tfrac{1}{2}mv^2 = 2mu(u + v) \approx 2muv, \tag{7.25}$$

if we assume $u \ll v$. The time between reflections is

$$\Delta t = \frac{d}{v}, \tag{7.26}$$

Fig. 7.11. Shock acceleration. A particle moving with velocity v in the laboratory frame is continuously reflected from two shocks moving symmetrically, at velocities $\pm u$ in the laboratory frame, toward each other. The particle is reflected elastically from each shock in the shock frame, and consequently is accelerated in the laboratory frame [equation (7.25)].

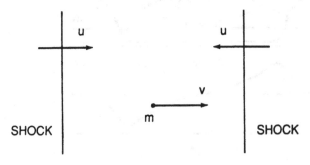

where d is the distance between the shocks (assumed here constant, since $u \ll v$). Equations (7.25) and (7.26) give the rate of energy gain by the particle:

$$\frac{dE}{dt} = \frac{2mv^2u}{d} = 4\left(\frac{u}{d}\right)E, \tag{7.27}$$

so that

$$E = E_0 \exp\left[4\left(\frac{u}{d}\right)t\right]. \tag{7.28}$$

Thus, given a suitable first-stage acceleration mechanism to provide an initial 'seed' population of particles with high enough velocity v (so that $v \gg u$), their energy will increase exponentially on a timescale $\tau = (4d/u)$, which, for sufficiently large u, can be of the order of seconds.

The situation referred to above, namely the symmetric motion of a single pair of shocks, is highly idealized. In a more realistic situation, such as an assembly of Petschek-type reconnection regions (Section 7.1.2), there will be a number of hydromagnetic shocks moving randomly in all directions, and the simple kinematic treatment above is replaced by a diffusion equation of the form

$$\frac{\partial f}{\partial t} = \frac{1}{p^2}\frac{\partial}{\partial p}\left[p^2 D(p)\frac{\partial f}{\partial p}\right], \tag{7.29}$$

where $f(p, t)$ is the momentum distribution function of the particles, and $D(p)$ is a diffusion coefficient whose form depends on the nature of the turbulent shock distribution. Under the approximation that the randomly moving fluid elements are hard spheres [cf. Ramaty *et al.* (1979)]

$$D(p) = \frac{1}{3}\left(\frac{u^2}{lv}\right)p^2, \tag{7.30}$$

where l is the mean free path of the particles (equivalent to d above). The solution to (7.29) with $D(p)$ given by (7.30) is in the form of a Bessel function, and this indeed seems to be a reasonable form for momentum spectra of interplanetary protons [see Ramaty (1986)].

7.1.8.3 *Particle acceleration and runaway*

Consider the equation of motion of an electron in a region of plasma in which there is an applied external electric field $\mathbf{E}$

$$m_e\dot{\mathbf{v}} = -e\mathbf{E} - v_c m_e \mathbf{v}, \tag{7.31}$$

where v_c is the electron–electron collision frequency, given by [cf. equations (4.68) and (4.69)]

$$v_c = \frac{12\pi e^4 n \ln \Lambda}{m_e^2 v^3}. \tag{7.32}$$

Considering only electric fields parallel to the particle velocity, we may take the scalar form of (7.31) and write

$$\dot{v} = \frac{e}{m_e} E - \frac{12\pi e^4 n \ln \Lambda}{m_e^2 v^2}. \tag{7.33}$$

For given n and E we can then solve equation (7.33) for $v(t)$. However, we may gain more immediate physical insight by examining the behavior of two electrons, one with initial velocity $v(0)$ small, and another with $v(0)$ large. For the former electron, as long as

$$v(0) < v_{\text{crit}} = \left(\frac{12\pi e^3 n \ln \Lambda}{m_e E} \right)^{1/2}, \tag{7.34}$$

the right-hand side of (7.33) is negative and the electron eventually decelerates to a stop. On the other hand, for $v(0) > v_{\text{crit}}$ the electron initially accelerates; this reduces the drag force still further and its acceleration increases. For large times the velocity can be made as large as we please (subject to relativistic considerations where applicable).

This phenomenon is known as *electron runaway*: electrons with an initial velocity above the threshold velocity efficiently accelerate and 'run away', while initially slow electrons come to rest. To gain an idea of what portion of an initial ensemble of electrons suffers runaway acceleration, we note that the threshold velocity equals the electron thermal velocity v_e for a field

$$E = E_D = \frac{12\pi e^3 n \ln \Lambda}{m_e v_e^2} = \frac{12\pi e^3 n \ln \Lambda}{kT}. \tag{7.35}$$

This field is known as the *Dreicer field*. From the definition of the Debye length [equation (4.14)], we may write

$$E_D = 3 \ln \Lambda \frac{e}{\lambda_D^2}. \tag{7.36}$$

The threshold velocity for an arbitrary applied field E can now be written

$$v_{\text{crit}} = v_e \left(\frac{E_D}{E} \right)^{1/2}, \tag{7.37}$$

and we see that large fields imply low v_{crit} and a large portion of runaway electrons. If the potential drop across the acceleration region is large compared with the thermal energy of the pre-acceleration distribution; i.e., if

$$e\phi = e \int E \, dl \gg kT, \tag{7.38}$$

then the process of electric field acceleration results in a nearly monoenergetic stream of electrons of energy $e\phi$, with the number of electrons in the stream given by integrating all the pre-acceleration Maxwellian distribution from v_{crit} to infinity. The number of electrons we can accelerate by Dreicer runaway depends on the rate of repopulation of electrons with $v > v_{\text{crit}}$ from electrons in the bulk distribution. Kruskal and Bernstein (1964) have provided formulae for this rate of electron runaway, viz.

$$n_{\text{runaway}} = 0.35 n v_{\text{c}} \left(\frac{E}{E_{\text{D}}} \right)^{-3/8} \exp\left[-\left(\frac{2E_{\text{D}}}{E} \right)^{1/2} - \frac{E_{\text{D}}}{4E} \right]. \tag{7.39}$$

Tsuneta (1985) used this result to show that for $E/E_{\text{D}} \gtrsim 0.3$ and for reasonable flare coronal parameters, Dreicer runaway can result in sufficient nonthermal electron flux to account for observed hard X-ray fluxes. However, the origin of such a high electric field is not addressed. Indeed, Holman (1985) showed that, because of the high inductance of solar flare loops (a consequence of their great size compared to laboratory circuits), the number of electrons that can be accelerated is limited by the constraint that the current be of the same order as the initial pre-flare current, associated with the curl of the magnetic field. He found that such a current of accelerated electrons can produce microwave bursts (Section 7.3) in solar flares, requiring an electric field $E/E_{\text{D}} \approx 0.02$–0.1. However, the number of electrons required to produce hard X-ray bursts (Section 7.2) corresponds to an unacceptably large current from a single region (and a much larger electric field). He therefore argued that many small (and therefore low-inductance) regions are required to produce the electrons responsible for hard X-ray bursts. Note also that the Kruskal and Bernstein result [equation (7.39)] was derived for the case of an infinite homogeneous plasma and ignores the decelerating effect of the charge separation expected by electron runaway out of a finite acceleration region, such as would apply to the solar flare problem. Thus, the efficiency of runaway acceleration in solar flares is still an open question.

We shall now discuss the main observational manifestations of the impulsive phase. We begin with the hard X-ray burst, considered by many as the most fundamental signature of this phase.

7.2   Hard X-ray bursts

Hard X-rays from solar flares were first detected by Peterson and Winckler (1959) using a balloon observatory. Since then, various spacecraft instruments have greatly extended our data base in this wavelength range. Not only has instrument sensitivity been greatly improved, but significant steps toward investigation of the spatial and polarization properties of the emitted X-ray radiation have recently been made.

7.2.1   *Intensity and spectra*

Figure 7.12 shows a typical large hard X-ray burst. The spiky structure of the burst on timescales of a few seconds is clearly evident; indeed, it has been proposed [e.g., de Jager and de Jonge (1978)] that each of these spikes represents a separate energy release (or 'elementary flare burst'). Hard X-ray bursts in flares range from simple isolated burst with a total duration of the order of seconds, to extended bursts lasting up to 10^3 s, such as the great flare of 4 August 1972 [see Hoyng *et al.* (1976)].

Figure 7.13 shows some typical photon energy spectra in the hard X-ray range – such spectra are generally quite steep, varying by several decades over a single decade in energy. In order to identify the characteristic photon energy associated with a given finite-bandwidth detector, the form of the

Fig. 7.12. A typical hard X-ray burst, observed with the Orbiting Solar Observatory 5 satellite. The flux versus time in three different channels is shown. Note the spiky nature of the burst in all channels and the steep fall-off of flux with energy (see Figure 7.13) (from Kane *et al.*, 1979*b*).

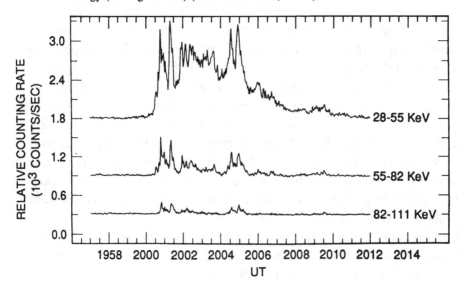

spectrum must be known and the emission in that detector channel appropriately weighed to determine the average photon energy observed. This leads to a somewhat involved iteration procedure (Dennis, 1982) whereby the points determine the spectrum and, in turn, the spectrum determines the points. For this reason it is in fact impossible to simply plot a series of intensity versus energy $[I(\varepsilon)]$ points – a parametric mathematical form for $I(\varepsilon)$ must be assumed and its parameters determined by iteration to self-consistency with the observed counts in each detector channel. A power-law function

$$I(\varepsilon) = a\varepsilon^{-\gamma} \tag{7.40}$$

Fig. 7.13. Hard X-ray spectra for four different events. Over the limited energy range shown here, the spectra are quite steep and are well (but not uniquely) represented by power law forms as shown. The horizontal bars represent detector channel widths and the vertical bars uncertainties in the count rate (from Kane *et al.*, 1979*b*).

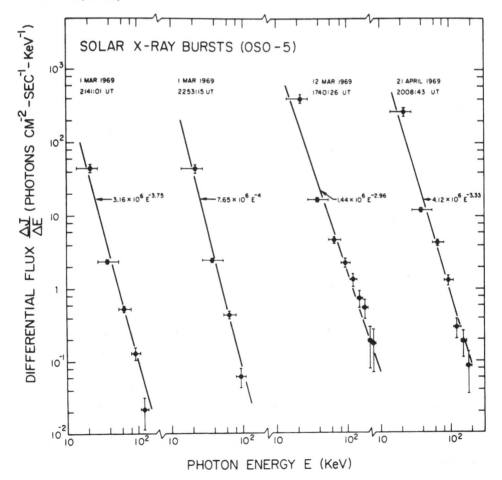

is a frequently used form in this procedure; values of γ typically run in the range from about 3 to 8, although recently (Lin *et al.*, 1984) a much steeper component ($\gamma \simeq 11$) at low energies ($\varepsilon \leqslant 20$ keV) has been observed using cooled germanium spectrometers, whose higher resolution allows a much more sensitive determination of spectral shape. Another canonical function used to fit hard X-ray spectra is that from a single temperature thermal (Bremsstrahlung) source [see equation (5.47)]:

$$I(\varepsilon) = \frac{b}{\varepsilon T^{1/2}} \exp(-\varepsilon/kT). \tag{7.41}$$

It has been pointed out [see Craig and Brown (1976)] that because of the 'filtering' property of the Bremsstrahlung cross-section [equation (5.11)], fine details in a Bremsstrahlung spectrum are 'washed out', leaving only general trends. Thus, simple fitting functions like (7.40) and (7.41) are, in a sense, the best we can do, given the limited spectral coverage of most detectors. One must bear in mind, however, that these are simply mathematical constructs designed to parametrically distinguish one spectrum from another, and should not be interpreted as precise functional forms that the spectra must satisfy.

7.2.2   Nonthermal models

In the case where the hard X-rays are presumed to be produced by a beam of suprathermal particles, impinging on a thin or thick target, we can use the form of the hard X-ray spectrum to infer the characteristics of the energetic electrons responsible for emitting the Bremsstrahlung. For example, if the hard X-rays are emitted throughout a thick target (Section 5.2.1.1) and have a power law spectral form (7.40), we can write, using equations (5.37) and (5.40), an expression for the power in energetic electrons above reference energy E_1

$$\mathscr{F}_1 = \frac{4\pi R^2 C}{\kappa_{\mathrm{BH}} \overline{Z^2}} \frac{\delta - 1}{B(\delta - 2, \frac{1}{2})} a E_1^{\,2-\delta}, \tag{7.42}$$

where $\delta = \gamma + 1$, by equation (5.36). At each instant, therefore, knowledge of the parameters $a(t)$ and $\gamma(t)$ allows determination of the electron spectral parameters $\mathscr{F}_1(t)$ and $\delta(t)$. The reference energy is frequently taken to be around 20 keV [see, e.g., Hoyng *et al.* (1976)], although we stress here that the choice of E_1 is arbitrary, as long as the power law form (7.40) applies at all energies $\varepsilon > E_1$.

The values of $\mathscr{F}_{20\,\mathrm{keV}}$ inferred from equation (7.42) can be very large, up to 10^{31} erg s^{-1} for large events [e.g., Tanaka and Zirin (1985)]. Indeed, the total energy in nonthermal electrons could be considerably higher than this if the spectrum extended to significantly lower values of E_1 [equation (7.42)]; for example, for $\delta = 6$ ($\gamma = 5$) the value of $\mathcal{F}_{10\,\mathrm{keV}}$ is 16 times the value of $\mathcal{F}_{20\,\mathrm{keV}}$. Inference of E_1 from, e.g., changes in the shape of the Bremsstrahlung spectrum, is therefore of paramount importance in flare studies. Unfortunately, at low energies, thermal Bremsstrahlung from the heated flare coronal plasma becomes significant, and the determination of the low-energy cutoff in the accelerated nonthermal electron spectrum must rely on subtle clues. Presently no rigorous bounds on E_1 have been derived.

If we therefore assume that a nonthermal thick-target Bremsstrahlung interpretation of solar flare hard X-ray emission is valid, such large energy fluxes place strong constraints on the flare energy-release process. The rate of energy dissipation corresponding even to an E_1 as high as 20 keV implies the complete annihilation every second of a 200-G field throughout a volume of 10^{27} cm^3. Such a rate of energy release is quite formidable, being much larger than the estimates in Section 7.1.4. This is further emphasized when one adds the requirement that the release be in the (low entropy) form of directed, accelerated particles rather than in the (higher entropy) form of bulk heating (Smith, 1980). It is largely for this reason that so much attention has been devoted to the study of hard X-ray bursts; an unambiguous verification of the thick-target nonthermal model of Bremsstrahlung production would place powerful constraints on the overall flare process.

7.2.3 *Thermal models*

As we shall see shortly, the same hard X-ray yield produced by a confined source of hot electrons is much less demanding energetically, since in principle 100 % of the electron energy can go into Bremsstrahlung, in itself a relatively small component of the flare energy budget. In fact a thermal Bremsstrahlung model was the first proposed to explain hard X-ray emission during flares (Chubb *et al.*, 1966). Furthermore the early available data consisted of spatially unresolved spectra, easily explainable (Brown, 1974) by a multi-temperature source [see equation (5.49) *et seq*], and the available observations on burst polarization and directivity were consistent with a thermal source (see below). There were, however, early objections to this interpretation. Kahler (1971a, b) pointed out that at the 10^8 K and upward temperatures required of such a thermal hard X-ray source, conductive cooling would be so efficient that a supply of energy even larger than that required for a nonthermal model would be necessary in order to maintain the source of the burst for the observed duration. He further pointed out that it would be exceedingly difficult to confine such a hot plasma because of the very long collisional mean free paths $\lambda \approx v/v_c \sim v^4$ [equation (7.32)] of the electrons and consequent high probability of escape. Therefore, despite the fact that the thermal model could not be ruled out on observational grounds, there seemed to be strong theoretical arguments against it.

The viability of a thermal hard X-ray source was restored when Brown *et al.* (1979) showed how Kahler's objections to a thermal source could be countered by appealing to the bulk plasma physics of such a source. A very large electrical current would be generated if, as a result of the long mean free paths of the electrons in the source, they all started streaming out of the source into the surrounding cooler plasma. The electric field set up by the resulting charge separation and by inductive effects would accelerate ambient electrons into a reverse current, as discussed in Chapter 6. Studies by Manheimer (1977), in connection with laboratory plasma fusion research, have shown (Section 4.5) that such a return current is unstable to the growth of ion-acoustic waves if the magnitude of the velocity of the associated drift exceeds a critical value

$$v_{crit} = \alpha \left(\frac{T_e}{T_i} \right) c_s, \tag{7.43}$$

where c_s is the ion-acoustic speed [equation (3.84)] and α is a function of the electron-to-ion temperature ratio [equation (4.114)] which has a value around unity for $T_e \gg T_i$. Since the ion-sound speed c_s is a factor $(m_p/m_e)^{1/2} \simeq 43$ times smaller than the electron thermal speed v_e, the streaming of hot electrons at their thermal speed will inevitably result in an unstable return current and the generation of ion-acoustic waves. With a high level of ion-acoustic waves present, the escaping electrons have an effective collision frequency which is significantly higher than in a wave-free regime. Their mean free path is substantially reduced, and their confinement is much easier to accomplish.

The model to emerge from these considerations has been termed the 'dissipative thermal model', and is shown schematically in Figure 7.14 where it is compared with the nonthermal thick-target model. The ion-acoustic wave fronts confining the hot Bremsstrahlung-producing electrons move along the flaring loop at a velocity v_{crit}; the bulk of the electrons are confined by these fronts, but very energetic electrons, with correspondingly long mean free paths, even in a wave-turbulent regime, are able to escape and penetrate to the footpoints of the loop, much as in the pure nonthermal model. The energetic efficiency of the model is determined by the escape of these electrons and by the anomalous conductive dissipation of the ion-acoustic fronts as they travel along the loop. The resulting efficiency is typically better than that of a thick-target model, although this depends on the parameters of the model, such as density, length, and temperature (Smith and Lilliequist, 1979; see below).

Fig. 7.14. Comparison of nonthermal (thick-target) and thermal models of hard X-ray production in solar flares. In the nonthermal model the energy release primarily accelerates deka-keV electrons which stream downward along guiding magnetic field lines and impact on the dense chromosphere, producing most of their Bremsstrahlung there. In the thermal model, the energy release principally heats electrons which free-stream outward, driving an unstable return current to preserve charge and current neutrality. The turbulent ion-acoustic wave regions ('fronts') so established move relatively slowly down the loop and confine all but the most energetic electrons at the top of the loop. The hard X-ray Bremsstrahlung signature of this model therefore, has two components, one in the corona and one in the chromospheric footpoint (after Emslie and Rust, 1979; see also Figures 7.15 and 7.16).

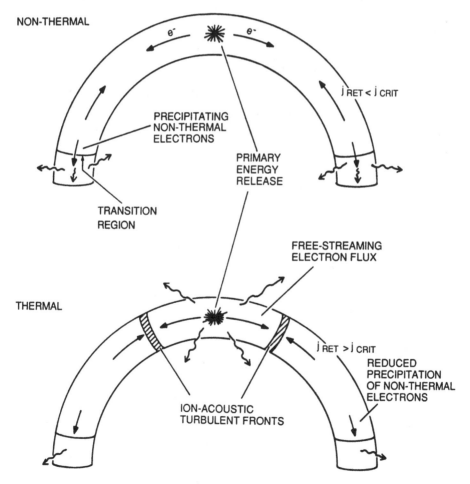

Before considering the detailed observational predictions of these two contrasting models, it is instructive to actually calculate their Bremsstrahlung efficiencies, defined as the ratio of ergs of Bremsstrahlung radiation produced to ergs of energy deposited in the source. Let us first consider the nonthermal thick-target model. From equations (5.15), (5.9), and (5.19), we find that the number of photons of energy ε (per unit ε) produced by the electron of energy E_0 in a collisional thick-target model with a pure hydrogen atmosphere is

$$m(\varepsilon, E_0) = \frac{8\alpha}{3} \frac{r_0^2}{C} \frac{m_e c^2}{\varepsilon} (E_0 - \varepsilon), \tag{7.44}$$

where we have neglected the logarithmic factor in the Bremsstrahlung cross-section. Using $C = 2\pi e^4 \ln \Lambda$ and $r_0 = e^2/m_e c^2$, we find that the total energy in emitted photons is

$$E_\gamma = \int_0^{E_0} \varepsilon m(\varepsilon, E_0) \, d\varepsilon = \frac{2\alpha}{3\pi \ln \Lambda} \frac{E_0^2}{m_e c^2}, \tag{7.45}$$

and the energetic efficiency of the process is

$$\eta = \frac{E_\gamma}{E_0} = \frac{2\alpha}{3\pi \ln \Lambda} \left(\frac{E_0}{m_e c^2} \right). \tag{7.46}$$

For E_0 in the hard X-ray range, e.g., 20 keV, this evaluates to $\eta \simeq 3 \times 10^{-6}$, showing that nearly all the energy of the injected electrons goes into heating ambient electrons (in mostly small-angle collisions; see Section 4.4) rather than actually producing Bremsstrahlung (in relatively large-angle collisions). This is the basic reason behind the somewhat embarrassing energy requirements of such models.

Now consider the 'dissipative thermal model' of Brown et al. (1979) discussed above. For a source of length L, cross-sectional area S, density n, and temperature T, the Bremsstrahlung emissivity is, by equation (5.45), neglecting factors of order unity,

$$I(\varepsilon) = \left(\frac{8}{\pi m_e k} \right)^{1/2} \frac{8\alpha}{3} r_0^2 m_e c^2 \frac{n^2 SL}{\varepsilon T^{1/2}} e^{-\varepsilon/kT} \quad \text{photons s}^{-1} \text{erg}^{-1}, \tag{7.47}$$

giving a total photon power output

$$P_\gamma = \int_0^\infty \varepsilon I(\varepsilon) \, d\varepsilon = \left(\frac{8k}{\pi m_e} \right)^{1/2} \frac{8\alpha}{3} r_0^2 m_e c^2 n^2 SL T^{1/2} \quad \text{erg s}^{-1}. \tag{7.48}$$

To sustain such a source in the presence of conductive cooling through motion of the ion-acoustic fronts requires an energy input equal to the increase in volume of the source per unit time multiplied by the energy density of the source, viz.

$$P_{in} = S\left(\frac{kT}{m_p}\right)^{1/2} \tfrac{3}{2} nkT, \qquad (7.49)$$

so that the efficiency is

$$\eta = \frac{P_\gamma}{P_{in}} = \left(\frac{8m_p}{\pi m_e}\right)^{1/2} \cdot \frac{16\alpha}{9} \cdot (nLr_0^2)\left(\frac{m_e c^2}{kT}\right)$$

$$= 4 \times 10^{-16} \frac{nL}{T}. \qquad (7.50)$$

Typical parameters of large hard X-ray bursts are emission measure $EM = (n^2 SL) = 10^{48}$ cm^{-3} and $T = 2 \times 10^8$ K [e.g., Crannell *et al.* (1978)]. To determine n we must assume a source area. Using the size of the Hα kernels (Chapter 1) as a guide, we put $S = 10^{17}$ cm^2. At a time when $L = 10^8$ cm, the source density must therefore be $(10^{48}/10^{17} \cdot 10^8)^{1/2} = 3 \times 10^{11}$ cm^{-3}, and the corresponding efficiency is, by equation (7.42), $\eta = 6 \times 10^{-5}$, which is small but still some 20 times more efficient than the thick-target model.

7.2.4 *Diagnostics*

The above discussion clearly shows that an understanding of the energetics of flare energy release requires that the scenario appropriate to the hard X-ray production be determined. Historically, the nonthermal thick-target model was the first to be developed in detail. Brown (1971) showed how the parameters of the injected electron distribution could be derived from those of the hard X-ray spectrum [see equations (5.36) and (5.37)]. A revision of these calculations, allowing for the effect of reverse current ohmic losses (Section 4.5) from the Bremsstrahlung-producing electrons, was performed by Emslie (1980). A significant result of this latter calculation is that there is an upper limit to the hard X-ray yield from such a model; in the (high flux) limit where ohmic losses dominate, a doubling (say) of the electron flux doubles the decelerating electric field [equation (4.88)], thereby halving the hard X-ray yield per electron [equation (5.15)]. Thus, the total hard X-ray yield is, in this limit, independent of the injected electron flux, while for low collision-dominated fluxes it is proportional to the injected electron flux. Calculations which take into consideration the velocity–space instability of a nonthermal beam of electrons interacting with a relatively cool background have only recently been carried out. McClements (1987*a*) found a further reduction in the hard X-ray yield because of noncollisional energy redistribution processes caused by the excitation of electrostatic Langmuir waves through the two-stream instability (Section 4.5).

The alternative ('dissipative thermal') model in its originally proposed form was mostly qualitative in nature and neglected several potentially significant physical processes. Various calculations, mostly numerical, have since been made to refine this model. These have dealt with the escape of high-energy tail particles from the thermal source (Vlahos and Papadopoulos, 1979; Smith and Brown, 1980), the microwave signature of the model (Emslie and Vlahos, 1980), and the evolution of the confining conduction fronts and their interaction with the ambient atmosphere (Smith and Auer, 1980; Smith and Harmony, 1982). A comparison of the radiation rise timescales of the model with observed time profiles of hard X-ray and microwave bursts has been carried out by Batchelor *et al.* (1985), who found generally good agreement. However, this result has been criticized (MacKinnon, 1985) on the grounds that Batchelor *et al.* equated the source expansion time to a *rise* time. Due to the cooling of the source, as the energy is redistributed through a larger and larger volume of plasma, the expansion time should rather correspond to the *decay* time of the hard X-ray burst. The agreement found by Batchelor *et al.* may thus simply be indicative of the relative symmetry of the time profile of X-ray bursts [see de Jager and de Jonge (1978)].

It is one of the pressing problems in solar flare physics today to determine to what extent each of the above models represents the actual situation in flares. Let us therefore discuss the expected hard X-ray signature of each model and its compatibility with available observations.

As with any radiation field, the characteristics of hard X-ray emission from solar flares are its intensity (erg cm^{-2} s^{-1} sr^{-1} Hz^{-1}), spectrum, anisotropy, polarization, and temporal structure. Therefore, as well as simply investigating the form of the photon energy spectrum $I(\varepsilon)$, we can ask for its dependence on viewing angle [i.e., the degree of anisotropy in $I(\varepsilon)$, or its directivity] or position within the flaring region (i.e., the spatial structure of the source), its degree of (linear) polarization, and the timescales associated with rise and fall of emission.

7.2.4.1   *Spatial structure*

The structure of Bremsstrahlung X-rays in both nonthermal thick-target and thermal models has been summarized by Emslie (1981, 1983). In a thick-target model, the emission at any point in the flare loop is roughly proportional to the flux of nonthermal electrons at that point times the ambient density; i.e.,

$$I(\varepsilon, N) = n \int_{\varepsilon}^{\infty} F(E, N)\sigma_{\mathrm{B}}(\varepsilon, E)\, \mathrm{d}E, \tag{7.51}$$

where ε is photon energy and N is column depth. We expect $I(\varepsilon, N)$ to show at first an increase with N as a result of the increase in n associated with greater depth in the atmosphere, and then, as we progress beyond the stopping distance for electrons of energy $E \geq \varepsilon$, viz. $N_{\mathrm{stop}} \approx \varepsilon^2/3C$ [equation (4.76)], to decline as a result of the attenuation of $F(E, N)$ through Coulomb collisions (Emslie and Machado, 1987).

In the dissipative thermal model, on the other hand, the emission per unit volume at a point s along a loop is given by the thermal Bremsstrahlung formula [equation (5.47)]

$$I(\varepsilon, s) = D\, \frac{[n(s)]^2}{\varepsilon[T(s)]^{1/2}} \exp[-\varepsilon/kT(s)], \tag{7.52}$$

which generally increases with both the density and temperature at the point in question.

In order to proceed further with a comparison of thick-target and thermal models we need to specify the density and temperature as a function of position in the loop [the column density N follows from $N = \int_0^s n(s')\, \mathrm{d}s'$]. Such an atmospheric model can either be computed empirically, using values consistent with observed emission measures, line ratios, etc., or by actually solving the hydrodynamic (or magnetohydrodynamic) equations for the atmospheric response to the flare-energy input. The latter approach has been adopted by Emslie *et al.* (1986a) and by Brown and Emslie (1987), who used the models of Emslie and Nagai (1985) in connection with thick-target electron heating (Section 8.1.1). Brown and Emslie also noted that the X-ray emission from the hot plasma produced by the electron heating can, for plausible parameters, be at least as important as the nonthermal Bremsstrahlung produced by the beam, particularly at photon energies $\varepsilon < 10\,\mathrm{keV}$. The above-mentioned results depend considerably on the parameters (beam flux, initial atmospheric structure, etc.) used in the hydrodynamic model. Therefore in order to simply highlight the differences between thick-target and thermal models, we reproduce Emslie's (1981) results in Figures 7.15 and 7.16, which use simple parametric constructs for the background density and temperature structures.

We see from these figures that the thick-target model is characterized by an initial 'footpoint' structure, which subsequently evolves into a filled loop as hydrodynamic expansion of the pre-flare chromospheric material drives material upward into the loop. By contrast, the thermal model initially

Fig. 7.15. Fractional Bremsstrahlung yield per kilometer of loop length versus depth in nonthermal (dashed lines) and thermal (solid lines) models, for various photon energies E. Note the predominance of footpoint emission at all energies in the nonthermal thick-target model, and the appearance of a strong coronal component, produced by electrons confined by the ion-acoustic turbulent fronts (Figure 7.14), in the thermal model. Note also the change of length scale for coronal and chromospheric regions; the column density scale is smooth along the entire figure (from Emslie, 1981).

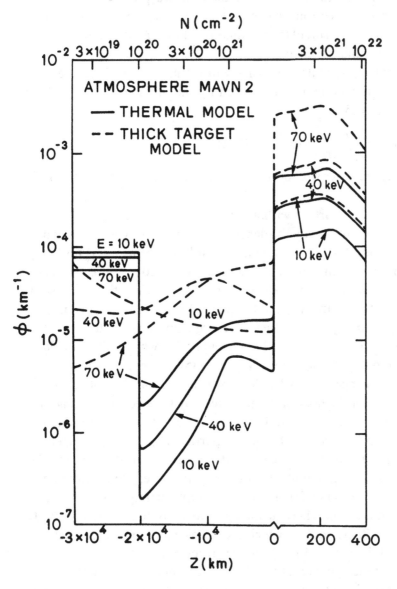

exhibits a bright central region in addition to the footpoint emission from the escaping high-energy electrons (Figure 7.14). This central region spreads downward along the loop as a consequence of the diffusion of heat into the surrounding plasma. Note that this is not an expansion of the source in the hydrodynamic sense.

These characteristics of the spatial emission are sufficiently diverse to afford a discriminating test between the two models, even though it is distinctly possible that some form of 'hybrid' model, involving characteristics of both models, applies [see, e.g., Emslie and Vlahos (1980)]. Nevertheless, it is worthwhile examining the available observations and seeing to what extent they favor either model.

Direct imaging of hard X-ray sources in solar flares has only been accomplished fairly recently with the Hard X-Ray Imaging Spectrometer (HXIS) instrument on the Solar Maximum Mission (SMM) spacecraft and with the Solar X-Ray Telescope (SXT) onboard the Hinotori satellite. Hoyng *et al.* (1981) reported images in the energy range 16–30 keV which

Fig. 7.16. Schematic of the spatial structure in hard X-rays, obtained by convoluting the results of Figure 7.15 with a broad spatial filter typical of current instrumentation. Nonthermal models can be expected to evolve from the low-density configuration into the high-density configuration as a result of 'evaporation' of heated chromospheric material (Chapter 8); this causes the coronal parts of the loop to brighten considerably in hard X-rays.

**MODEL N-T
LOW DENSITY CORONA**

**MODEL N-T
HIGH DENSITY CORONA**

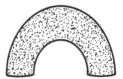

MODEL T

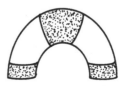

change from a 'double footpoint' structure early in the event to a more amorphous structure later. This observation may represent a change from a nonthermal source into a thermal one, but an alternative interpretation (Brown and Emslie, 1987) is that the chromospheric material heated by the electron beam expands and flows upward into the upper part of the loop (see Chapter 8), carrying the Bremsstrahlung target protons with it. Subsequent observations using the SMM HXIS instrument [e.g., Duijveman *et al.* (1982); Machado *et al.* (1985)] are also subject to the same ambiguity of interpretation. The situation is even further confused by the SXT images which, in apparent direct contrast to those from HXIS, show predominantly coronal sources in the same energy range during limb events (Tsuneta *et al.*, 1984*a*,*b*). This may be due to the different methods by which the imaging data are gathered by the two instruments: the HXIS instrument transmits data for the brightest pixels only, and it is possible that a large diffuse component in the corona could be completely ignored (MacKinnon *et al.*, 1986). Also, the quantity of (optically thin) emitting material along the line-of-sight in, e.g., a semi-circular loop varies with position in the loop, tending to produce 'footpoints' in disk center events, as opposed to a predominance of high-level emission in limb events (see MacKinnon *et al.*, 1986; also Figure 7.16). More statistically significant hard X-ray imaging data, at significantly higher photon energies (> 50 keV, say, to avoid the problem of 'contamination' from the high-emission-measure thermal sources created by the electron beam itself through atmospheric heating) and preferably with finer resolution (say, ≈ 1 arc sec, the collisional stopping length of a 10-keV electron in a plasma of density 10^{11} cm^{-3}), are required to significantly address this issue of spatial structure (see Epilogue).

Through a set of fortuitous circumstances, Kane *et al.* (1979*a*) reported on a flare that was simultaneously observed by two detectors, one on the International Sun–Earth Explorer 3 (ISEE 3) satellite at the inner Lagrangian point of the Sun–Earth system and one on the Pioneer Venus Orbiter (PVO) satellite around the planet Venus. Although neither detector had direct imaging capability, the location of the flare was such that although PVO had an unobstructed view of the flaring region, ISEE 3, at a different solar longitude, would only see the upper parts of the flaring loop ($h \geqslant 25\,000$ km), the lower parts being occulted by the western limb of the Sun. This provided a crude 'imaging' of the flare into upper and lower components. Kane *et al.* found that the ratio of fluxes observed by the two detectors implied that only $\frac{1}{600}$ of the emission came from the upper parts of the loop visible from ISEE 3. This measurement was at energies of a few hundred kiloelectron volts, well above the energies which could be 'contaminated' by thermal emission from the 10^7 K plasma created by the flare heating. Brown *et al.* (1981) have interpreted these observations in terms of a simple collision-dominated thick-target model, explaining both the ratio of fluxes and the different shapes of the two spectra observed. At the top the spectrum is that of a thin target with $\gamma = \delta + 1$ [equation (5.31)], while over the whole loop the spectrum is that of a thick target with $\gamma = \delta - 1$ [equation (5.36)].

Rust *et al.* (1985) have observed the propagation of X-ray emission along large loops with lengths from around 50 000 to 250 000 km. The velocity of the enhancements is of order 1000 km s^{-1}, too fast for hydrodynamic motions which propagate at a speed comparable to the ion thermal speed. However, the observed velocity *is* consistent with the velocity associated with an electron thermal conduction front which, in the nonclassical regime appropriate to this study, can propagate at a significant fraction of the electron thermal speed [e.g., Campbell (1984)]. Rust *et al.* interpreted the observation as evidence for thermal wave fronts in solar flares. However, Machado *et al.* (1985, 1988*b*) have pointed out that propagation of non-thermal electrons and hydromagnetic shocks from footpoint to footpoint may also play a significant role in these phenomena.

7.2.4.2 *Directivity*

The cross-section for Bremsstrahlung production is anisotropic because it is a function not only of the directions of the ingoing and outgoing electrons and of the outgoing photon (especially at relativistic energies), but also of whether the emitted photon is parallel or perpendicular to the collision plane [Gluckstern and Hull (1953); cf. Chapter 5]. For a thermal source with its randomized velocity distribution, any preference for direction of the emitted photons is lost, and the source emits isotropically. On the other hand, for a source consisting of accelerated nonthermal electrons propagating along guiding magnetic field lines and interacting with an ambient medium, there is a preferential direction for the ingoing electrons, and consequently for the outgoing photons. Consequently, such a source emits Bremsstrahlung anisotropically. Calculations of the expected anisotropy, or directivity, of the hard X-ray Bremsstrahlung from a thick-target model have been carried out by Brown (1972) and Leach and Petrosian (1983). Brown showed that quite high directivities should be present in a thick-target model, with flares observed perpendicularly to the guiding magnetic field lines appearing considerably brighter than those observed along the field lines. For vertical magnetic fields, this implies that hard X-ray events on the limb should be systematically brighter than events observed at disk center. Leach and Petrosian (1983), however, included the effects of pitch angle diffusion in their calculations (whereas Brown had simply dealt with mean values of quantities) and showed that the predicted directivity was not as great as suggested by Brown, although still significant and potentially observable.

Calculations of the directivity from a thermal source have been carried out by Emslie and Brown (1980). Any isotropic thermal source will, by symmetry, emit radiation isotropically. However, the presence of temperature gradients in the source causes preferential electron drift in the direction toward lower temperatures, giving rise to a skewed electron phase-space velocity distribution of the form (Manheimer, 1977)

$$f(v) = \frac{n_e}{(2\pi v_e^2)^{3/2}} \left\{ \exp\left[-\frac{(v_\parallel - u) + v_\perp^2}{2v_e^2} \right] \right\} \left\{ 1 + \frac{F(v_\parallel - u)}{3n_e m_e v_e^4} \left[\left(\frac{v}{v_e}\right)^2 - 3 \right] \right\},$$

(7.53)

where v_e is the thermal velocity, $v_\parallel$ and $v_\perp$ are the parallel and perpendicular velocity components, u is a bulk drift velocity, and F is the heat flux. Setting $u = F = 0$ we find the usual isotropic Maxwellian distribution, but for $u = 0$ and finite F, the distribution is asymmetric, with a high-energy tail on the positive $v_\parallel$ direction and a current-neutralizing slow bulk drift in the negative $v_\parallel$ direction (see Figure 7.17). [For large negative values of v in equation (7.4) the function $f(v)$ becomes slightly negative. However, for such velocities we may replace $f(v)$ by zero without significantly affecting the normalization and predicted directivity (Chambe and Hénoux, 1979).]

Fig. 7.17. Electron phase-space distribution function parallel to the guiding field lines in the thermal model of Figure 7.14. The high-energy tail in the positive $v_\parallel$ direction, produced by streaming of high-energy electrons out of the heated region, is balanced by a slow drift of the bulk distribution at velocity v_d. If v_d becomes too large, we generate ion-acoustic turbulence; this occurs at the boundary of the source region (see Figure 7.14).

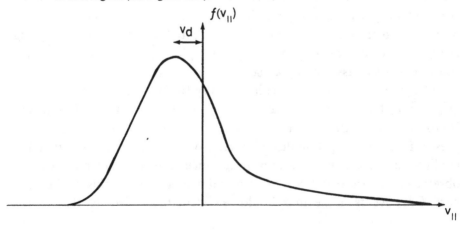

This skewed distribution gives rise to a finite directivity (and polarization; see Section 7.2.4.3) of the Bremsstrahlung emission. The calculations of Emslie and Brown (1980) place this directivity at about 10% with a predicted limb *darkening*, in contrast to the limb brightening of nonthermal models. Physically this arises because the thermal source is principally at the top of a loop and therefore on near horizontal field lines, in contrast to the vertical field lines assumed in the thick-target modeling of Brown (1972).

The simple pictures above are complicated by possible, and indeed probable, inhomogeneities in the hard X-ray source and by the fact that a substantial portion of the hard X-rays observed at the Earth are photons that are Compton backscattered from the solar photosphere. This albedo emission induces a further directivity [e.g., Hénoux (1975); Bai and Ramaty (1978)] and must be considered when a comparison with observations is made.

Observationally, directivity studies are of two types, statistical and stereoscopic. Datlowe *et al.* (1977) statistically examined a sample of several hundred hard X-ray flares and searched for a systematic limb brightening (or darkening) of events. Although their results regarding brightness variation are too noisy to be conclusive at a 2σ confidence level, they do show a systematic trend for limb flares to have harder spectra, indicating that emission of high-energy photons is directed to a significant degree and in a manner consistent with the theoretical predictions of the thick-target model.

On the other hand, Kane *et al.* (1980) obtained direct information on the anisotropy of the hard X-ray emission in a series of events observed simultaneously with the ISEE 3 and PVO satellites, while the PVO was orbiting Venus and situated at a different heliocentric longitude than the Earth. Their results show no systematic trend toward limb brightening or darkening.

In summary, present results, both theoretical and observational, are inconclusive concerning the directivity of hard X-ray Bremsstrahlung in flares. Nevertheless, such measurements have a significant potential for testing models of Bremsstrahlung production and energy release in flares, and we must look forward (Epilogue) to better measurements and more refined theoretical predictions in this area.

7.2.4.3 *Polarization*

The remarks of the preceding paragraph also hold for the current status of observation versus theoretical prediction as to the degree of polarization expected in hard X-ray bursts. Polarization of hard X-ray emission, like directivity, has its physical origin in the difference between the Bremsstrahlung cross-sections for emission in and perpendicular to the plane defined by the incoming and outgoing electron. Again, for an isotropized thermal source, the integrated polarization is zero, since there is no preferred plane, while, for a beamed nonthermal source, a non-zero polarization is expected.

Theoretical calculations of the polarization from a thick-target model have been carried out by Brown (1972) and Leach and Petrosian (1983). As with the directivity results, the inclusion of pitch angle diffusion in the calculations significantly lowers the expected polarization from that predicted using only mean values. Emslie and Brown (1980) showed that moderate polarization, again of a few percent, also results from a thermal source with a skewed velocity distribution function (7.53) arising from the presence of temperature gradients. Hénoux (1975) and Bai and Ramaty (1978) calculated the contribution to the polarization from photospherically backscattered photons and found that this process induces polarization into primarily unpolarized (primary) sources, while it can also reduce the polarization from highly polarized primary sources.

Observations of hard X-ray polarization in flares are relatively sparse. In the early 1970s the Soviet series of *Interkosmos* satellites obtained the first crude and somewhat inconclusive measurements of Bremsstrahlung polarization [see, e.g., Tindo and Somov (1978) and references therein]. In 1981, results from the hard X-ray polarimeter on Spacelab 1 showed detectable polarization significant at the 2σ-level (Tramiel *et al.*, 1984). However, a comparison of these data with theoretical models (Leach *et al.*, 1985) was inconclusive in discriminating between nonthermal and thermal interpretations, largely due to the relatively low photon energies observed and to the fact that most of the Bremsstrahlung emitted by a nonthermal electron beam is produced only after the beam has undergone considerable isotropization through Coulomb collisions with ambient electrons and protons. Leach *et al.* suggested that measurement of the polarization signal from the collisionally thin upper part of the loop alone (avoiding the diluting weakly polarized component produced by the near-isotropized electrons near the feet) would be a powerful discriminator. Direct imaging of this coronal component seems a remote possibility at the level of current instrumentation, but it is possible that measurement of a flare situated just behind the solar limb [cf. Kane *et al.* (1979a)] could provide the necessary occultation of the chromospheric component [Chanan *et al.* (1988); see Epilogue].

7.2.4.4 *Temporal structure*

As can be seen in Figure 7.18, hard X-ray bursts exhibit a rather spiky time structure (Dennis, 1985). Fourier analyses of time profiles of the flux of hard X-ray bursts reveal no evidence for any single characteristic timescale (Hoyng *et al.*, 1976; Kiplinger, 1986). However, it is fairly clear that the release of energy, as exhibited by the hard X-ray flux, is not smooth. De Jager and de Jonge (1978) suggested that hard X-ray bursts are composed of a superposition of fundamental 'building blocks', which they named *elementary flare bursts* (EFBs), each with a duration of the order of 5–10 s. Observations with 10 ms time resolution using the SMM Hard X-Ray Burst Spectrometer (HXRBS) (Kiplinger *et al.*, 1983) show clear evidence for time structure down to tens of milliseconds (see also similar observations in microwaves; Section 7.3.2). This represents a powerful constraint on source models, since it represents either the collisional stopping time (in the case of nonthermal models) or the conductive cooling time (in the case of thermal models). The collisional stopping time for an electron of energy E is [equation (7.32)]

$$\tau_c = v_c^{-1} = \frac{2 \times 10^8 E^{3/2}}{n}, \tag{7.54}$$

Fig. 7.18. A hard X-ray burst on 6 June 1980 observed at 10 ms time resolution with the SMM Hard X-Ray Burst Spectrometer. The error bars are one standard deviation. Note the statistically significant burst between 23:34:45.5 and 23:34:46.0 UT, with rise and fall times of the order of 10–20 ms (courtesy B. Dennis and A. Kiplinger).

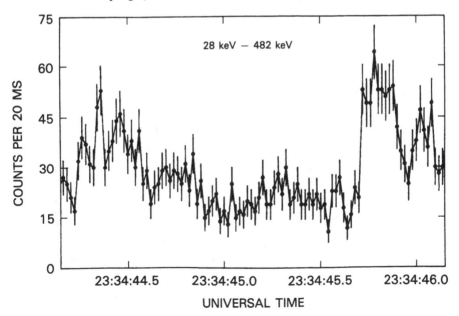

with E in kiloelectron volts, while the conductive cooling time is (cf. Section 7.2.3)

$$\tau_{cond} = L/c_s = \frac{L}{10^4 T^{1/2}} \tag{7.55}$$

[equation (3.84)]. For these times to be sufficiently short quite stringent constraints are imposed on source densities and sizes; i.e., $n \gtrsim 3 \times 10^{11}$ cm^{-3} and/or $L \lesssim 20$ km [see Kiplinger *et al.* (1983) for details]. However, it must be pointed out that it is by no means clear that all hard X-ray bursts can be modeled as a superposition of such ultrafast structures in the same way as EFBs.

7.2.5   Summary

We have seen that hard X-ray bursts are a very important part of the flare phenomenon in that they may require a dominant portion of the total flare energy to produce the energetic electrons that in turn produce the observed Bremsstrahlung. This conclusion depends, however, on the characteristics of the Bremsstrahlung-producing electrons. The question is whether the electrons form a high-energy component interacting with a cooler background plasma, or form part of a bulk-energized 'thermal' population. Numerous observational tests to discriminate between these scenarios and to address the fundamental nature of the energy release process have been devised. However, due to the limitations of current instrumentation, the question remains open, and we can but look forward (see Epilogue) to more precise observational constraints that future generations of instruments can provide.

7.3   Microwave bursts

Another powerful diagnostic of energetic electrons is the radio emission they produce when, for example, individual electrons spiral around magnetic field lines (gyrosynchrotron radiation) or when electrons act together through collective plasma processes. For typical solar flare conditions, the gyrofrequency and plasma frequency are comparable in the gigahertz or microwave range. In this section we shall discuss the physical mechanisms responsible for the microwave emission in flares and their characteristic observational signatures. Then we shall address microwave observations of flares, notably those at high spatial resolution using large interferometric arrays. Excellent review of microwave observations of flares can be found in Kundu and Vlahos (1982) and Kundu (1983*a*).

7.3.1   Emission mechanisms

Microwave radiation in solar flares arises from three distinct processes (see Chapter 5), i.e., thermal Bremsstrahlung, gyrosynchrotron radiation, and collective plasma processes. Thermal Bremsstrahlung is simply free–free continuum emission, the low-frequency (high-wavelength) end of the blackbody radiation curve for a hot gas at temperature T. At frequencies v such that $hv \ll kT$, the classical limit is a very good approximation and we can replace the usual Planck formula

$$B_v(T) = \frac{2hv^3}{c^2} \frac{1}{\exp(hv/kT) - 1}$$

by the classical Rayleigh–Jeans formula (2.4′) or

$$B_v(T) = \frac{2kT}{c^2} v^2. \tag{7.56}$$

We see that this radiation, which is optically thick and unpolarized, has a spectrum that increases like v^2 with the slope of this spectrum determining the temperature T of the source. For any radio source, thermal or nonthermal, we can define a *brightness temperature* T_b by inverting (7.56):

$$T_b = \frac{c^2 B_v(T)}{2kv^2}. \tag{7.57}$$

For nonthermal sources T_b may be a function of v and may have little or no bearing on the actual temperature in the source.

The presence of a magnetic field introduces another important source of microwave emission, namely gyrosynchrotron radiation. Low-energy electrons gyrating around magnetic field lines emit circularly polarized radiation at the gyrofrequency $v_B = eB/2\pi m_e c = 2.8 \times 10^6 B$ [equation (4.21)]. For mildly relativistic electrons, however, the radiation is beamed significantly, resulting in a radiation signal from the source which is more 'spiky' than a simple harmonic wave. As discussed in Section 5.3, Fourier decomposition of this 'spiky' signal into its components yields power at harmonics as high as 100, by which time the spectrum has almost blended into a continuum. Since the degree of beaming is a function of the energy of the emitting electron, the decomposition into harmonics is a complicated function of the electron energy, and the resulting microwave spectrum is quite complicated; see equation (5.65). Dulk and Marsh (1982) gave formulae for the emission and absorption coefficients for various electron energy distributions, such as Maxwellians and power laws, and derived the expected forms of microwave spectra. In the limit of low frequencies, the source is optically thick and typically has a positive spectral slope ranging from v^2 for a thermal distribution [see equation (5.70)] to about v^3 for a power law distribution of nonthermal electrons. At higher frequencies, the source becomes optically thin, and the spectrum, which now reflects the emission coefficient ε_v rather than the source function $S_v = \varepsilon_v/\kappa_v$, changes to a negative slope; v^{-8} for a thermal population and between v^{-1} and v^{-4} for nonthermal populations, depending on the power law index δ of the electrons (see Figure 7.19). The maximum power is emitted at frequencies corresponding to optical depth of order unity; this gives a formula for v_{max},

Fig. 7.19. Gyrosynchrotron microwave flux spectra for nonthermal (both relativistic and non-relativistic) and thermal electron distributions, after Dulk and Marsh (1982) who used Ginzburg and Syrovatskii's (1965) curves for relativistic power law distributions. δ is the power law index of the electron flux spectrum. Note (i) the appearance of a maximum in S at a frequency v_{max}, which is a function of the parameters of the source region [equation (7.58)] and (ii) the characteristically different spectra from nonthermal and thermal (single temperature) distributions.

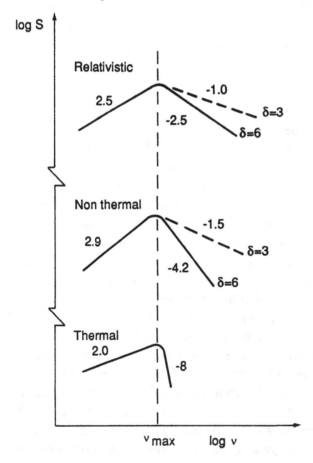

the frequency at which maximum power is emitted, as given by equation (5.69) or

$$v_{max} = \begin{cases} A_1 N^{1/3} B^{2/3}, & \text{nonthermal}, \\ A_2 T^{1/2} B, & \text{thermal}, \end{cases} \tag{7.58}$$

where N is the column density of the source, and A_1 and A_2 are constants. Thus, measurement of the 'turnover' frequency v_{max} gives a measurement of one of the parameters N(or T) and B, if an independent estimate can be made of the other one.

Microwave emission can also be produced by coherent oscillations of plasma; see Section 5.4. As pointed out in Section 3.4.1, a plasma, when excited, resonates at an angular frequency [equation (3.75)] $\omega_{pe} = (4\pi n e^2/m_e)^{1/2}$. A possible mechanism for causing growth of such plasma waves (Langmuir waves) to significant energy levels is the passage of a beam of high-energy electrons, which gives rise to a positive slope in the electron velocity distribution function (see Section 4.5). These plasma waves can couple nonlinearly to produce an electromagnetic wave at the second harmonic, $\omega = 2\omega_{pe}$ (Tsytovich, 1970; Smith and Spicer, 1979), which, if not strongly gyroresonance absorbed, can propagate to the observer. Emslie and Smith (1984) and McClements (1987b) calculated the expected microwave energy from this process and found that it can possibly exceed the gyrosynchrotron contribution. This raises the possibility that microwave emission in solar flares is predominantly due to such coherent Langmuir plasma oscillations.

Another collective process which may provide significant microwave emission is the maser (*M*icrowave *A*mplification by the *S*timulated *E*mission of *R*adiation) process. Melrose and Dulk (1982) pointed out that electrons reflected in a magnetic mirror configuration (Section 4.3) have a distribution with a 'loss cone' corresponding to the particles with low pitch angles that have escaped from the trap and that therefore have not been reflected. Since nature abhors a vacuum (even one in velocity space), electron–cyclotron waves are set up to scatter electrons into this loss cone. Melrose (1980) and Melrose and Dulk (1982) gave details of the growth rate of the electron–cyclotron waves. These waves can couple to produce microwave emission, and the amount of microwave power produced by this process can be very large. Since gyroresonance absorption in the source is very likely, due to the nature of the emission process, Melrose and Dulk (1982) concluded that transport of such electromagnetic waves, which occurs both along and perpendicular to the field lines, could be a significant factor in the overall flare energy budget.

7.3.2   *Observations*

The ability of observations at radio wavelengths to resolve fine-scale detail (< 1 arc sec) is currently unequaled at any other wavelength range. This is due to: (a) the steadiness of the Earth's atmosphere on size scales of a radio wavelength, which allows theoretical resolving powers of telescopes to be achieved (a limit unattainable with optical telescopes much larger than a few inches in aperture because of atmospheric 'seeing'), and (b) the recent advent of large arrays of radio telescopes which provides baselines of several kilometers, thereby bringing the theoretical resolution angle to subarc sec values. Consequently, considerable detail in microwave images has been observed, notably with the Very Large Array (VLA) which can provide images and spectra simultaneously [cf. Marsh and Hurford (1982)].

Superposition of VLA microwave maps of flares onto Hα images of the flare shows (Kundu, 1983b) that the microwave emission at 6 cm typically, but not always, comes from a region between the two footpoints, presumably near the top of the flaring loop, i.e., near the presumed site of magnetic dissipation that is responsible for the energy release. Marsh and Hurford (1980) studied several cases of microwave bursts at 2-cm wavelength. In the impulsive phase the emission was dominated by a compact source located between two Hα kernels. Later the microwave source was larger and elongated along the magnetic field lines, presumably connecting the Hα kernels, see Figure 7.20. Further observations at 2-cm wavelength (Marsh *et al.*, 1981) revealed emission divided into two components, one on each side of the magnetic neutral line, with opposite senses of circular polarization, and Shevgaonkar and Kundu (1985) reported that the 2-cm radiation frequently comes from two footpoint sources at the base of the loop. These studies provide powerful evidence for the acceleration of nonthermal electrons down both legs of the loop, the different senses of polarization being due to the different sense of the longitudinal magnetic field component when measured from the top of the loop downward.

Observations at 20 cm [e.g., Melozzi *et al.* (1985)] generally show extended sources separated from the shorter wavelength emission (Velusamy *et al.*, 1987) and are believed to be gyrosynchrotron radiation from the hot thermal plasma contained within the loop. A clear picture of the various mechanisms responsible for microwave emission at various wavelengths requires accurate determination of the emission at various wavelengths simultaneously, observations which have been relatively rare but which promise to become more plentiful in the near future (Epilogue).

The strength of gyrosynchrotron radiation is proportional to the

Fig. 7.20. Microwave source structure at 15 GHz (2 cm) (white contours) on an Hα map of the flare during the impulsive phase (*a*) and gradual phase (*b*). The sources are clearly resolved (see beam width contour at lower right) and show the evolution of the microwave source, along the loop connecting the Hα footpoints, as the flare proceeds (from Marsh and Hurford, 1980).

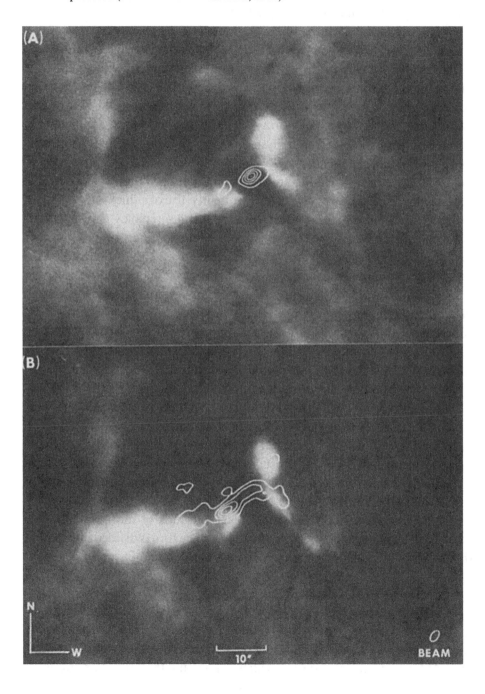

magnetic field strength, and it might be expected that the emission from the loop footpoints, where the magnetic field strength is largest, would predominate. However, as pointed out by Petrosian (1981) and Dulk and Marsh (1982), gyrosynchrotron emissivity is also highly peaked in a direction perpendicular to the magnetic field lines, so that when looking at right-angles to the field at the top of the loop we see a greater fraction of the emission than when looking down along the field lines near the footpoint.

The number of electrons required to produce the microwave burst was first thought to be considerably (up to 3–4 orders of magnitude) less than the corresponding number of electrons, of similar energy, required to produce hard X-ray bursts [e.g., Takakura (1969, 1975)]. However, Gary (1985) pointed out that these earlier analyses assumed thin-target emission for the hard X-rays, and the inefficiency of this model compared to a thick-target model (cf. Section 7.2) causes an overestimate of the number of electrons required for the hard X-ray burst. Consequently, use of a thick-target model for the hard X-ray production can remove this discrepancy entirely.

Kaufmann *et al.* (1984) observed very fast time structure, down to a few tens of milliseconds, in microwave observations at 22 and 44 GHz. (Similar rapid fluctuations in hard X-rays have also been reported; see Section 7.2.4.4.) This is evidence that the impulsive phase of solar flares is composed of a great number of separate energy releases, a picture qualitatively consistent with the magnetic reconnection scenarios presented in Section 7.1.1. Just how the microwave flux can rise (and, more significantly, fall) on such short timescales is not at present clearly understood.

This very fast time structure was first noted by Dröge and Riemann (1961), Elgarøy (1961, 1962), and de Groot (1962) at lower frequencies. The word *spike*, first used by de Groot, is now a common radio-astronomical term for these short-lived radiobursts from flares. An excellent review is given by Benz (1987). Spikes most frequently appear simultaneously with Type III radio bursts, which are strong radio signatures of electron beams in the corona, generated in the flare-energy release. No other flare radiation is so fragmented: the spikes consist of ten thousands of individual events, which may indicate that the electron acceleration in turn is divided into tens of thousands of individual elements, each providing a microflare (Benz, 1985).

7.4   Gamma-ray bursts

Gamma-ray lines in flares are produced by nuclear reactions involving accelerated protons and light nuclei; see Section 5.7. Gamma-ray spectroscopy was first achieved using the Gamma-Ray Spectrometer (GRS) on SMM, which recorded the intensities of lines and their time variations. It was discovered (Forrest and Chupp, 1983; Kane *et al.*, 1986) that γ-ray bursts are an impulsive phase phenomenon, time synchronized with the hard X-ray burst to within a second or so. This result demonstrates that acceleration of million electron volt electrons and protons is a direct product of the magnetic reconnection process and not a second-stage phenomenon, as has previously been advocated (see Section 7.1.8.2). Gamma-ray observations thus provide valuable diagnostics of the primary energy-release process.

Excellent reviews of γ-ray processes in solar flares have been presented by Ramaty *et al.* (1975, 1979) and Ramaty (1986). In this book we shall concentrate our discussion on observational results and their implications for conditions in the impulsive phase.

7.4.1 *Intensity and spectra*

Expressions for the γ-ray yield resulting from the injection of a beam of high-energy protons into a solar plasma, with appropriate elemental abundances [cf. Allen (1973)], have been given by Ramaty *et al.* (1975) and Ramaty (1986). These authors present results for various injected energy spectra and for both thin- and thick-target interaction models (Section 7.2.2). The spectrum is composed of both prompt and delayed features (see Section 5.7, where prompt features like the nuclear de-excitation lines at 4.4 MeV from ^{12}C and at 6.1 MeV from ^{16}O were discussed). The most notable delayed feature is the 2.223 MeV deuterium formation line caused by neutron capture on hydrogen; see equation (5.86):

$$^1\mathrm{H} + {}^1\mathrm{n} \rightarrow {}^2\mathrm{H} + \gamma. \tag{7.59}$$

This line is not produced until the neutron (resulting from a collision of an accelerated proton on a heavy nucleus) has slowed sufficiently for the cross-section for reaction (7.59) (which is a decreasing function of the neutron velocity) to become sufficiently large. This line is consequently produced deep within the solar atmosphere and should be limb-darkened, as is observed (Ramaty, 1986, and references therein).

Ramaty *et al.* (1979) calculated synthetic prompt γ-ray spectra using a thick-target model of proton interaction with the solar atmosphere and photospheric element abundances. Figure 7.21 compares such a synthetic spectrum with γ-ray observations obtained during a flare. The excellent agreement between theory and observation is evident; detailed model fitting of observations has led to the conclusion that the abundances of carbon and oxygen in the chromospheric flare site are lower (relative to iron) than in the photosphere.

7.4.2   *Implications for particle acceleration*

The number of energetic protons required to produce the observed γ-ray flux has been calculated, and one finds that the energy content of these protons can be quite large [of order 10^{30} erg in protons above 1 MeV (Ramaty, 1986)]. Although these protons produce negligible X-ray Bremsstrahlung, they can be important for heating the deep chromosphere in flares (Machado *et al.*, 1978; Emslie, 1983).

Observations made with the HXRBS and GRS instruments show that the hard X-ray and γ-ray fluxes peak within about a second of each other (the microwave flux also peaks at this time). This result (Forrest and Chupp,

Fig. 7.21. Comparison of observations of a γ-ray event of 27 April 1981 with a theoretical spectrum calculated from a thick-target proton bombardment model. Note the appearance of the prompt de-excitation lines (notably of ^{12}C and ^{16}O), as well as the delayed deuterium formation line at 2.223 MeV (labeled '*n*'). The elemental abundances have been adjusted to provide the best fit to the observations; this fitting procedure implies that, for example, the abundances of C and O are increased from their photospheric value relative to Fe (courtesy E. L. Chupp).

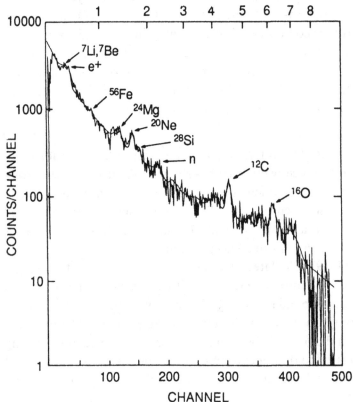

1983; Kane *et al.*, 1986) again shows that protons are accelerated promptly as a direct result of the magnetic dissipation process and not in a 'second-stage' phenomenon. This conclusion is further strengthened by observations (Chupp *et al.*, 1982) of high-energy *neutrons* from flares. Because of their relatively prompt arrival time at the GRS detector, these neutrons must have velocities close to the speed of light and consequently very high energies (up to 600 MeV), and they must have been produced within seconds of the γ-ray peak. Such neutrons can only be produced by nuclear reactions involving protons of energies in excess of 1 GeV. We therefore see that the γ-ray and neutron data from the GRS demand that prompt acceleration of large fluxes of protons, with energies up to 1 GeV, is a requirement of the flare-energy release. Recently, Bai *et al.* (1983) proposed that protons are accelerated as part of a second-step acceleration process, involving first-order Fermi acceleration [e.g., Jokipii (1966); see Section 7.1.8.2] between a pair of upward moving shock fronts. This second-step acceleration process is also, according to Bai and Ramaty (1978), responsible for accelerating electrons of a few hundred kiloelectron volts, and its slight delay relative to the impulsive phase energization may explain the observed delay of very high-energy hard X-rays ($\varepsilon \gtrsim 150$ keV) relative to deka-keV hard X-rays [e.g., Hoyng *et al.* (1976)].

7.5 Impulsive EUV bursts

EUV radiation in the solar atmosphere originates in layers with temperatures below about 10^6 K, i.e., in the transition region and upper chromosphere. As a result of energy transport throughout the flaring atmosphere, dense, relatively cool pre-flare chromospheric layers are impulsively heated to temperatures sufficient to produce EUV radiation. The enhanced densities in these newly heated regions cause the EUV emission measure to increase, producing a burst of emission. As the heating proceeds, deeper and deeper layers are progressively heated enough to produce EUV emission. It is therefore important to realize that the EUV-emitting plasma in flares is not a fixed source with changing parameters, but rather a phase surface that in general moves downward throughout the impulsive phase, bringing ever denser regions into play.

7.5.1 *Observations*

Early observations of EUV bursts in flares were made by studying the ionospheric response to arriving EUV radiation (Donnelly, 1976). When the ionosphere responds to a burst of EUV radiation, radio waves reflected from ionospheric layers suffer sudden, impulsive Doppler shifts, leading to so-called *sudden frequency deviations* (SFDs). Using a suitable model of the Earth's atmosphere, observations of SFDs may be used to infer the incoming EUV flux. This method gives absolute fluxes to an accuracy of only about a factor of 4, but shows relative variations of the flux at the 1% level. The method is sensitive to EUV radiation in the waveband 10–1030 Å, corresponding to continua and lines emitted in the upper chromosphere.

Kane and Donnelly (1971), Donnelly and Kane (1978), and Emslie *et al.* (1978) showed that the broadband 10–1030-Å EUV flux inferred from SFD measurements peaks simultaneously with the hard X-ray burst, to within the timing accuracy of the data. Donnelly and Kane (1978) further showed that the rise time of the EUV burst corresponds most closely with the rise time of the hard X-ray channels around 30 keV. From this they concluded that electrons with this energy were responsible for exciting the EUV burst. From a center-to-limb study of EUV to hard X-ray ratios, they suggested that the EUV source consisted of a 'trench', or well, dug out of the solar atmosphere by high-energy electron heating, surrounded by a shallower well, presumably corresponding to soft X-ray irradiation (Machado, 1978; Somov, 1981; see Section 8.1.5).

EUV bursts have also been measured in narrow emission lines, formed at transition region temperatures. Notable among these is the O V line at 1371 Å, formed at temperatures around 2×10^5 K, since this line has been studied extensively with the Ultraviolet Spectrometer and Polarimeter (UVSP) instrument on the SMM. Emission in this line is also observed to peak synchronously with the hard X-ray burst [Cheng *et al.* (1985); Orwig and Woodgate (1986)]. These authors also showed that the portions of the EUV continuum, formed in the deep chromosphere near the temperature minimum [e.g., Vernazza *et al.* (1976)], are also observed to peak simultaneously with the transition region line emission and with the hard X-ray burst. This suggests that radiative transport is a significant mechanism for the energization of the lower atmosphere in flares, a point to which we shall return in Section 8.1.5.

7.5.2   *Implications for energy transport models*

The synchronism of the broadband EUV and X-ray bursts is indicative of a very fast energy transport mechanism between the energy-release site and the chromosphere, such as by high-energy electrons. The rise time studies of Donnelly and Kane (1978) support this conclusion. Yet Emslie *et al.* (1978) found that the ratio of intensities of EUV and hard X-ray bursts is incompatible with a thick-target electron heating model. Specifically, they found that such a model should produce far more 10–1030-

Å radiation than observed. This can be explained if either: (*a*) only a portion of the electrons producing the hard X-rays manages to reach the chromosphere (the rest being mirrored in a trap geometry; cf. Section 4.2), or (*b*) the area of injection is very small ($\approx 10^{16}$ cm^2) so that the high beam fluxes necessary to produce the observed hard X-rays over such a small area ablate much of the chromosphere to coronal temperatures (see Section 8.1.1), resulting in only a small fraction of the beam energy being deposited in the cool EUV-emitting chromospheric plasma. Such a model, when used in conjunction with a simple optically thin formula for the radiative losses in the chromosphere gives rise to a relatively simple (power-law) dependence between the hard X-ray and broadband EUV fluxes from event to event (Emslie *et al.*, 1978), a dependence which is in agreement with observations (Kane and Donnelly, 1971). More sophisticated calculations involving optically thick radiative transfer of the EUV photons, notably those forming the Lyman continuum, lead to conclusions still in accord with this behavior (McClymont and Canfield, 1986).

The synchronism of EUV line emission with hard X-ray flux is less straightforward. Energization of a loop by flare energy input does lower the atmospheric level of EUV-emitting plasma, thereby increasing the density, but it also tends to steepen temperature gradients through the presence of a larger conductive flux $F_c = KT^{5/2}\,dT/ds$ (Spitzer, 1962). Thus, the differential emission measure per unit cross-sectional area (see Section 2.6)

$$Q(T) = n^2 \frac{ds}{dT},$$

which, when integrated over the temperature range corresponding to formation of the species responsible for the spectral line under observation, gives a quantity proportional to the line intensity [equation (2.52)], need not necessarily rise in response to flare heating. Poland *et al.* (1984), Emslie and Nagai (1985), and Mariska and Poland (1985) studied this problem and found that only relatively brief bursts of electron heating can produce the observed rise and fall of EUV line fluxes. Other forms of heating (such as by thermal conduction) or large-duration bursts (larger than 5–10 s; cf. EFBs – Section 7.2.4.4) do not produce the required time profile, either through driving the $Q(T)$ *downward* initially or through a hysteresis caused by the large relaxation times after a prolonged period of heating.

7.6   Other impulsive signatures

We shall defer discussion of signatures relating to atmospheric response to the impulsive heating until Chapter 8. In this section we shall briefly mention some other signatures of impulsive flare energization that relate directly to the primary energy release.

7.6.1   *Iron Kα radiation*

Kα lines are formed when an electron in the K-shell of an atom is removed either by collisional ionization or photo-ionization, and a $2p$ electron de-excites into the $1s$ vacancy created by the inner shell ionization; see Section 2.2.4. Due to the relatively large abundance (compared to other metals) of neutral or near-neutral iron atoms in the solar chromosphere, a measurable Kα signal is expected during the impulsive phase of a solar flare when either collisional ionization by electron beams or photo-ionization by soft X-rays is present. The threshold energy for inner shell ionization of iron is 7.11 keV, and the Kα radiation takes the form of a closely spaced doublet at 6.39 and 6.40 keV, corresponding to de-excitation from the $^2P_{1/2}(K\alpha_2)$ and $^2P_{3/2}(K\alpha_1)$ levels respectively.

Statistically significant peaks in Kα intensity have been observed simultaneously with peaks in hard X-ray intensity (Tanaka *et al.*, 1983; Emslie *et al.*, 1986*b*). The intensities of these Kα bursts agree well with the intensities predicted by a thick-target model (Section 7.2.2), whose electrons cause the initial inner shell ionization of the iron atoms when they penetrate into the cool chromospheric layers. In addition to these spikes, one observes a generally dominant, slowly varying component which is interpreted as being due to photo-ionizational excitation [i.e., fluorescence; Bai (1979)] (Parmar *et al.*, 1984).

7.6.2   *Type III radio bursts*

Type III radio bursts are intense bursts of unpolarized radiation that show a rapid decrease in frequency with time [cf. Kundu (1965)]. They are interpreted as being plasma radiation at either the fundamental or second harmonic of the plasma frequency (Section 3.4.1) produced when Langmuir waves (plasma oscillations) scatter off ion-acoustic waves or other Langmuir waves respectively. The rapid decrease in frequency is then associated with a rapid decrease in the density of the excited plasma, and the conventional interpretation of Type III radio bursts is a high-energy electron beam, moving at an appreciable fraction of the speed of light and propagating outward from the flare site into ever less dense coronal regions. With a knowledge of the density versus height structure, $n(z)$, of the corona and with a presumption as to whether the radiation is fundamental or second harmonic, the velocity of the exciting beam can in fact be calculated from the frequency drift rate, viz. $dz/dt = (dz/dn)(dn/dv)(dv/dt)$.

Sturrock (1964) recognized a problem associated with an understanding of Type III bursts in pointing out that any distribution in velocities of the exciting electron beam would cause fast electrons to arrive at a given point in space before slower moving electrons. As a result the velocity distribution would have a positive slope $\partial f / \partial v$ and be unstable to the generation of Langmuir waves [equation (4.111)]. Quasi-linear relaxation of the electron distribution under the influence of these waves would cause the beam to stop in a very short distance and time, of the order of a kilometer and a nanosecond, respectively. These values are obviously much too small to account for observed Type III behavior. The resolution of this dilemma requires a careful balance between stabilizing the beam, in order that it can propagate, and the generation of enough plasma waves to account for the observed emission.

LaRosa (1988) considered a model in which the initial instability gives rise to a level of Langmuir waves which then saturate due to nonlinear wave–wave interactions. He found that the region of strong wave growth can be considered as a 'front', moving outward at somewhat less than the velocity of the electrons which generate the waves, and this front stimulates the radio emission. Electrons overtaking the front are diffused by it and generate more waves ahead of the front which, accordingly, move outward to the new region of wave growth. The region behind the front contains remnant plasma waves, which may continue to produce appreciable radiation.

The observed synchronism of Type III emission with the onset of other impulsive phase signatures, such as hard X-ray bursts (Švestka, 1976), is evidence for electron acceleration during the primary energy release. However, these electrons are ejected along open field lines into interplanetary space and contribute little to the hard X-ray emission, since the number of electrons involved is typically 4–5 orders of magnitude less than the number required to produce the hard X-ray burst by Bremsstrahlung (Lin, 1974). These electrons may have been accelerated on closed field lines and subsequently drifted onto open field lines through a magnetic field curvature drift [Emslie and Vlahos (1980); see equation (4.31)]. Occasionally, 'U' type bursts are seen in which the observed radiation drifts first downward and then upward in frequency as time proceeds (Tang and Moore, 1982). These bursts presumably correspond to electrons injected onto closed field lines.

While Type III bursts point to electron acceleration in the impulsive phase of solar flares, we note that Type III bursts are also observed in the absence of any visible flare, even though ejecta can be seen (Trottet *et al.*, 1984). We must conclude, therefore, that the acceleration of electrons can take place in magnetic flux tubes due to some perturbation or instability that is not sufficiently strong to lead to a visible flare, but that can cause chromospheric activity observable in Hα for example (Chiuderi-Drago *et al.*, 1986). Another way of saying this is that flares occur on all energy scales, and flare-like manifestations may be observed in both big and small flares, as well as in subflares and maybe bright points (Section 6.6). Chiuderi-Drago *et al.* followed an idea of Leroy and Mangeney (1984) to envision a scenario in which chromospheric material is ejected by changes in the magnetic field configuration, probably due to emerging flux. The ejected material acts as a piston to generate a shock which – if it travels perpendicularly to the magnetic field – effectively converts its energy into kinetic energy of accelerated particles [for details, see Leroy and Mangeney (1984)]. These accelerated particles will, in turn, produce the observed Type III bursts.

7.6.3   *Particle ejection (e, p, n)*

Flares not only eject Type III-burst-producing electrons into interplanetary space, but also eject energetic protons and neutrons (Section 7.4.2). The spectrum of these ejected particles can be directly measured *in situ* using spacecraft in heliocentric orbits. Although there is a strong correlation between observation of interplanetary particle events and solar flares, the correspondence is not perfect – some flares apparently do not emit substantial particle fluxes. Conversely, not all particle events have flares with which they obviously can be associated (Potter *et al.*, 1980).

Charged particles spiral along magnetic field lines, and, due to the Archimedean spiral-type structure of the interplanetary magnetic field (caused by the solar rotation), a connection with terrestrial monitors is only possible for events near the west limb of the Sun. While electron spectra typically display positive slopes in some velocity range, in keeping with their Type III-burst-producing properties, proton spectra typically have a power law form with fairly hard spectral indices of order 2 (van Hollebeke *et al.*, 1975). However, because of the strong influence of interplanetary electromagnetic fields on the particles as they propagate from Sun to Earth, it is difficult to learn much from these measurements about the spectrum of the particles when they are accelerated. Nevertheless, they do show that substantial fluxes of particles are accelerated in the flare process, a conclusion backed up by the neutron observations (Section 7.4.2). A point worthy of note is the hazard to astronauts from bombardment by these ejected particles; one of the largest proton flares ever recorded happened on 4 August 1972, just a year before the launch of the Skylab observatory with its Apollo Telescope Mount (ATM) package of solar instruments. At first sight, it might be considered unfortunate that Skylab just missed this notable flare, but, on reflection, it is doubtful whether the astronauts could have survived it (see also Rust, 1982).

As well as elementary particles, flares also accelerate quantities of heavier nuclei, such as ^{2}H, ^{3}He, and ^{4}He. A long-standing problem is to understand why the abundance of ^{3}He is preferentially enhanced in flare particle streams, while that of ^{2}H is suppressed. Two schools of thought exist: according to one, the flare site itself is somehow ^{3}He enriched; according to the other, the ^{3}He is preferentially accelerated over ^{2}H (Fisk, 1978). Kahler *et al.* (1987) studied a number of ^{3}He events, and their findings are consistent with a scenario in which ^{3}He particles are accelerated in the corona high above that part of the flare plasma which is responsible for the normally observed Hα and X-ray emissions. A clear interpretation of the magnetic field configuration in the accelerating area and of the different abundance anomalies would be a valuable asset in understanding the particle acceleration mechanism (or mechanisms) in the impulsive phase of flares.

8

The gradual phase

After a storm comes a calm

Commentaries, Acts 9, M. Henry (1662–1714)

After its initial abrupt release in the impulsive phase of a flare, the energy is transported to other regions of the atmosphere, often as it changes form. New areas of the atmosphere are affected, mainly due to heating, and it is this interplay of energy transport and atmospheric response that we refer to as the gradual phase of the flare.

It is therefore often considered that to a first approximation the life history of a flare consists of an impulsive phase, characterized by mainly nonthermal emissions (hard X-rays, γ-rays, radio waves, and neutrons) and a gradual phase characterized by predominantly thermal emissions (soft X-rays and UV and optical radiation). Since an entropy increase is inevitably associated with the thermalization of the flare particles and the equilibration of the initially created radiation fields, information is progressively being lost as the flare manifestations progress. Nevertheless, very important information concerning the flare plasma and the solar atmosphere can be obtained during the gradual phase. While the initial impulsive phase is over in a matter of minutes or less, the gradual phase can go on for many hours. Before high-energy observations became available certain aspects of the gradual phase were studied, and the flare concept from that time period corresponds to our gradual phase concept.

Figure 8.1 shows an example of a compact flare observed in soft X-rays and in a UV line ($\lambda = 1356$ Å) of Fe XXI, which here define the gradual phase, as compared with the same flare observed in hard X-rays, which define the impulsive phase. The impulsive phase, with its two hard X-ray bursts, was over by 1920 UT, while the gradual phase was then just underway and reached maximum intensity significantly later. This was not a large flare (M1.2, B1; see Section 1.2), but we notice that even in such a case there are flare emissions, like the UV line of O V, that do not easily fall into an either/or category as far as impulsive versus gradual phase is concerned. The O V emission shows the impulsive burst behavior with its two peaks in

Figure 8.1, but the emission stays on well into the gradual phase. We have discussed in the introduction to Chapter 7 why and how we subdivide the flare into its impulsive and gradual phases.

8.1   Atmospheric response

We shall now investigate how the solar atmosphere reacts to the energy that has been released cataclysmically in the impulsive phase within a magnetic loop in the corona and which is now being transported and transformed on its path throughout the flaring region.

8.1.1   *Collisional heating*

In Section 4.5 we saw how energetic, nonthermal electrons can be accelerated at the initial flare site in the upper, coronal part of magnetic loops and subsequently be degraded through Coulomb collisions in the

Fig. 8.1. Comparison of the time profiles of soft X-ray (1–8 Å) emission, hard X-ray (photon energy $\geqslant 29\,\text{keV}$) emission, and emissions in the 1371-Å line of O V ($T \approx 2 \times 10^5$ K) and the 1354-Å line of Fe XXI ($T \approx 1 \times 10^7$ K) from before onset of the flare of 1 November 1980, through the impulsive phase, until well after the maximum of the gradual phase (from Tandberg-Hanssen *et al.*, 1984).

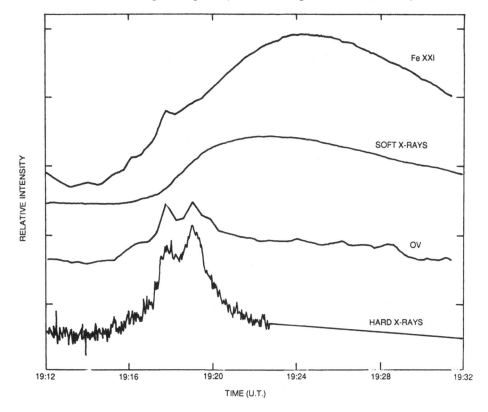

target atmosphere, a process referred to as thick-target electron bombardment.

As the electrons decelerate, two phenomena will take place. First, the braking of the electrons results in intense, short-wave Bremsstrahlung. These hard X-ray bursts have been treated in Section 7.2. Second, the deposition of the electrons will lead to heating of the atmospheric layer. Let us quantitatively evaluate the form of the atmospheric heating rate. For simplicity, we shall consider a flare loop with a uniform guiding magnetic field, thereby neglecting mirroring effects (Section 4.2), and we shall neglect collective effects (Section 4.5). Let us also consider the injection of an ensemble of suprathermal electrons with initial energy spectrum

$$
F_0(E_0) = \begin{cases} (\delta - 2)\dfrac{\mathscr{F}}{E_c^2}\left(\dfrac{E_0}{E_c}\right)^{-\delta}, & \text{for } E \geqslant E_c \\ 0, & \text{for } E < E_c, \end{cases}
\tag{8.1}
$$

where $\mathscr{F}$ (erg cm^{-2} s^{-1}) is the total injected energy flux above the cutoff energy E_c. The form (8.1) is for mathematical convenience and is not necessarily intended to correspond to a physically realistic spectrum. From continuity of electron flux along the direction of the field lines

$$
\mu F(E)\, \mathrm{d}E = \mu_0 F_0(E_0)\, \mathrm{d}E_0,
\tag{8.2}
$$

where E is the electron energy at column depth N [equation (4.70)], and μ_0 is the cosine of the pitch angle at injection. The heating rate at depth N is

$$
I_B(N) = n \int_E^c F(E)\left|\frac{\mathrm{d}E}{\mathrm{d}N}\right| \mathrm{d}E \quad \text{erg cm}^{-3}\,\text{s}^{-1}.
\tag{8.3}
$$

The integral may be transformed to one over E_0 by using (8.2), and the expression for $\mathrm{d}E/\mathrm{d}N$ has been calculated previously in (4.77). With these substitutions, the heating rate becomes

$$
I_B(N) = Cn \int_{E_0}^c \frac{F_0(E_0)\, \mathrm{d}E_0}{E_0\left(1 - \dfrac{3CN}{\mu_0 E_0^2}\right)^{2/3}}.
\tag{8.4}
$$

The integral limits depend on the atmospheric depth under consideration. For low N, such that electrons of energy E_c have not yet stopped [equation (4.76)], the cutoff energy E_c is the appropriate lower limit. However, for larger N, the minimum injected energy required to reach this depth is [equation (4.76)]

$$
E_{0,\min} = \left(\frac{3CN}{\mu_0}\right)^{1/2},
\tag{8.5}
$$

and the overlying atmosphere effectively imposes its own 'cutoff' energy on the relevant part of the injected energy spectrum. With the appropriate integral limits thus determined, substitution of the injected spectrum (8.1) into (8.4) gives

$$I_B(N) = (\delta - 2) \frac{Cn}{\mu_0} E_c^{\delta-2} \mathscr{F}$$

$$\times \int_{\max(E_c, [3CN/\mu_0]^{1/2})}^{\infty} \frac{E_0^{-\delta} \, dE_0}{E_0 \left(1 - \frac{3CN}{\mu_0 E_0^2}\right)^{2/3}}. \qquad (8.6)$$

To evaluate the integral we change the variable to $x = (3CN/\mu_0 E_0^2)$ and find

$$I_B(N) = \tfrac{1}{2} Cn(\delta - 2) \frac{\mathscr{F}}{E_c^2} B_x \left(\frac{\delta}{2}, \frac{1}{3}\right) \left(\frac{3CN}{\mu_0 E_c^2}\right)^{-\delta/2}, \qquad (8.7)$$

where B_x is the (incomplete) beta function:

$$B_x(\xi, \eta) = \int_0^x y^{\xi-1}(1 - y)^{\eta-1} \, dy \qquad (8.8)$$

and

$$x = \min\left(\frac{3CN}{\mu_0 E_c^2}, 1\right). \qquad (8.9)$$

For very low values of N, the incomplete beta function (8.8) evaluates to

$$B_x\left(\frac{\delta}{2}, \frac{1}{3}\right) \sim \frac{2}{\delta} \left(\frac{3CN}{\mu_0 E_c^2}\right)^{\delta/2}, \qquad (8.10)$$

so that the heating rate *per particle* $[J_B(N) = I_B(N)/n]$ is

$$J_B(N = 0) \sim C \frac{(\delta - 2)}{\delta} \frac{\mathscr{F}}{E_c^2}. \qquad (8.11)$$

The heating rate is independent of N, since no appreciable modification to the electron spectrum has resulted. For large values of N, the beta function becomes complete and

$$J_B(N) = \tfrac{1}{2} C(\delta - 2) \frac{\mathscr{F}}{E_c^2} B\left(\frac{\delta}{2}, \frac{1}{3}\right) \left(\frac{3CN}{\mu_0 E_c^2}\right)^{-\delta/2}; \quad N \geqslant \frac{\mu_0 E_c^2}{3C}. \qquad (8.12)$$

Since $J_B(N)$ is a decreasing function in this range, the maximum occurs at the lower limit, and is given by

$$J_{B,\max} = \tfrac{1}{2} C(\delta - 2) \frac{\mathscr{F}}{E_c^2} B\left(\frac{\delta}{2}, \frac{1}{3}\right); \quad N = \frac{\mu_0 E_c^2}{3C}. \qquad (8.13)$$

This is larger by a factor $(\delta/2) B(\delta/2, \tfrac{1}{3})$ than the value in (8.11) for $N = 0$, and we see that the form of $J_B(N)$ is such that $J_B(N)$ initially rises, as the electrons in the beam slow down and have their energy deposition rates correspondingly increased [equation (4.71)]. This trend continues until $N = N_c = \mu_0 E_c^2 / 3C$, beyond which point we start to lose electrons from the beam and the beam attenuation factor becomes dominant, causing $J_B(N)$ to decrease. The form of $J_B(N)$ for various spectral indices δ is shown in Figure 8.2.

Heating by energetic proton bombardment can be dealt with in a similar way (Emslie, 1978).

8.1.2   *Ohmic heating by reverse current*

The injection of a large flux of nonthermal electrons into a plasma very quickly sets up a charge separation and a large current. Both of these effects establish, the first electrostatically and the second inductively, an electromotive force which drives a beam-neutralizing return current. There is currently some debate about the relative importance of the roles of electrostatic and inductive processes in establishing the return current (Brown and Bingham, 1984; Spicer and Sudan, 1984). However, it is generally agreed that once established, the return current constitutes a relatively slow bulk drift of electrons in the opposite direction to the beam, and its dynamics have been discussed in Section 4.5. For sufficiently large injected fluxes, the heating associated with this return current $(I_B = \mathbf{j} \cdot \mathbf{E} = \eta j^2)$ will dominate over direct collisional losses of the beam. Near the injection point we have

$$I_B(N = 0) = \eta_0 T_c^{-3/2} e^2 F^2 \quad \text{erg cm}^{-3} \text{s}^{-1}, \qquad (8.14)$$

where $\eta_0 = 1.5 \times 10^{-7}$ is in cgs units [equation (4.83)], T_c is the coronal temperature, and $F = j/e$ is the beam flux in electrons cm^{-2} s^{-1}, related to the energy flux $\mathscr{F}$ by

$$F = \frac{(\delta - 2)}{(\delta - 1)} \frac{\mathscr{F}}{E_c}$$

[cf. equation (8.1)]. Comparison of (8.14) with (8.11) shows that the ratio of ohmic heating by the reverse current to collisional heating by the beam is

$$\beta = \frac{\delta(\delta - 2)}{(\delta - 1)^2} \left(\frac{\eta_0 e^2}{C} \right) \cdot \frac{\mathscr{F}}{n T_c^{3/2}}, \qquad (8.15)$$

so that $\beta > 1$ for sufficiently large $\mathscr{F}$. However, it is uncertain whether values of $\mathscr{F}$, sufficiently large to cause β to be $\gg 1$, can exist without implying a reverse current drift velocity high enough to cause ion-acoustic instability (Section 4.5).

Equations (8.11) and (8.14) are useful in deriving the temperature to which the corona can be heated during flares (Emslie, 1985). Assuming

Fig. 8.2. Normalized heating rate per particle versus dimensionless column depth variable x for collisional electron beam heating. Curves for two different δ (spectral index of the injected electron beam) are shown. Note the fairly localized maximum at $x = 1$; for $x < 1$ the heating rises with x as the electrons are slowed and consequently lose energy at a greater rate [equation (4.67)]; for $x > 1$ the heating falls as a consequence of attenuation of the beam flux by collisions.

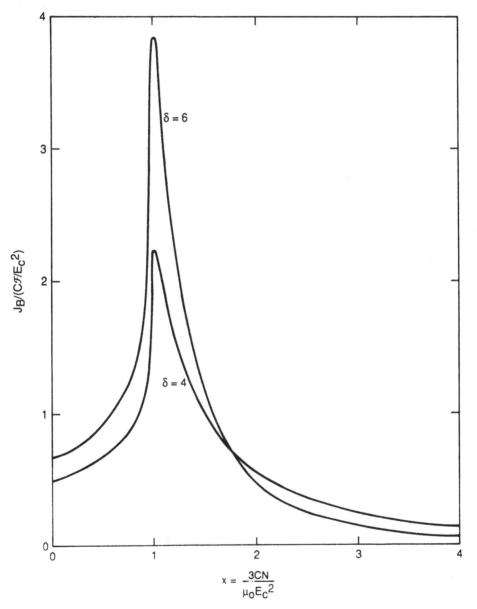

classical conductivity and neglecting radiative losses (a justifiable assumption for coronal densities and temperatures during flares, as may be verified *a posteriori*), we write the steady state energy equation for the coronal plasma at $N = 0$ in the form

$$I_{\text{B}}(N = 0) \approx \frac{K T_{\text{e}}^{7/2}}{L^2}, \tag{8.16}$$

where K is the classical coefficient of thermal conductivity and L is the length of the flare loop. For collisional heating, (8.11) and (8.15) yield

$$T_{\text{c}} \approx \left[\frac{C}{K} \frac{(\delta - 2)}{\delta} \frac{n L^2}{E_{\text{c}}^2} \right]^{2/7} \mathscr{F}^{2/7}, \tag{8.17}$$

while for ohmic heating associated with the reverse current, (8.14) and (8.15) yield

$$T_{\text{c}} \approx \left[\frac{\eta_0 e^2}{K} \frac{(\delta - 2)^2}{(\delta - 1)^2} \frac{L^2}{E_{\text{c}}^2} \right]^{1/5} \mathscr{F}^{2/5}. \tag{8.18}$$

These results are quite insensitive to the flare loop parameters n and L, and to the beam parameters $\mathscr{F}$, E_{c}, and δ. With $n = 10^{11}$ cm^{-3} and $L = 10^9$ cm, $E_{\text{c}} = 10$ keV $= 1.6 \times 10^{-8}$ erg, and $\delta = 6$, equations (8.17) and (8.18) give

$$T_{\text{c,collisional}} \approx 2 \times 10^4 \mathscr{F}^{2/7}, \tag{8.19a}$$

and

$$T_{\text{c,ohmic}} \approx 600 \mathscr{F}^{2/5}. \tag{8.19b}$$

For $\mathscr{F} = 10^{11}$ erg cm^{-2} s^{-1} (cf. Section 7.2 with a flare area of order 10^{18} cm^2), the temperatures in equations (8.19) both evaluate to about 2×10^7 K. This prediction is testable by measuring intensities of soft X-ray lines and, as we shall see below, is in good agreement with observations. Equations (8.19) also predict that T_{c} should vary like $\mathscr{F}^{2/7}$ in small (collision-dominated) events and like $\mathscr{F}^{2/5}$ in large (reverse current dominated) events. However, this prediction has yet to be tested observationally, due to the lack of reliable measurements of the temperature structure of flaring coronal plasma over a wide variety of flare sizes.

8.1.3   *Atmospheric response to beam heating*

The heating of chromospheric plasma in the legs of the loop by electron bombardment has been studied in considerable detail (Brown, 1973; Lin and Hudson, 1976; MacNeice *et al.*, 1984; Nagai and Emslie, 1984; Emslie and Nagai, 1985; Fisher *et al.*, 1985a, b; Fisher, 1987). It generally results in an upward, expanding motion of the chromospheric plasma, referred to by the – physically somewhat misleading, but generally accepted – expression of 'chromospheric evaporation' (Neupert, 1968;

Hudson, 1973; Antiochos and Sturrock, 1978). This process may be more or less violent, depending on the energy-flux level of the nonthermal electron heating. Below a certain threshold of the energy flux the chromosphere responds in a fairly passive way; i.e., we observe a 'gentle evaporation' (Fisher *et al.*, 1985*b*; Schmieder *et al.*, 1987) driven by thermal conduction from the upper hot plasma heated by the nonthermal electrons, a process similar to the one first suggested by Carmichael (1964) and pursued by Sturrock (1966) and Kopp and Pneuman (1976). If the energy flux exceeds this threshold, which depends on the chromospheric structure and the energy spectrum of the electrons, the atmosphere is unable to radiate away the deposited energy. The form of the radiative loss function $\phi(T)$ for an optically thin plasma is shown in Figure 8.3. We see that in the region from a few times 10^5 K to about 10^7 K, the radiative losses generally decrease with temperature, a result which arises physically from the decreasing contribution of line emission as more and more species are ionized. Above 10^7 K, the free–free continuum (Bremsstrahlung; Section 5.2.1) contribution comes into significant play and the radiative losses start to rise again, like $T^{1/2}$ (see Section 5.2.1.2). If the flare energy input at a given atmospheric level exceeds a value corresponding to the radiative losses $n^2\phi(T)$ at a temperature $T \approx 10^5$ K, the peak in the $\phi(T)$ curve, then the imbalance of input power and radiative losses leads to an increase in the temperature of the plasma. At a higher temperature, the radiative power of

Fig. 8.3. Radiative loss function $\phi(T)$ as a function of temperature. Solid line shows Raymond *et al.*'s (1977) results and the dotted line the analytic fit due to Rosner *et al.* (1978). Previous work by Pottasch (1965) and McWhirter *et al.* (1975) gave similar, albeit somewhat lower, values for $\phi(T)$.

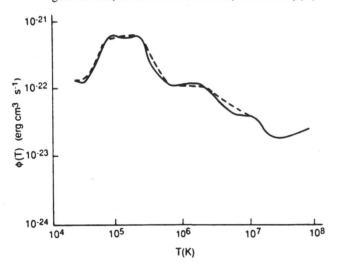

the plasma is, however, reduced and so an even greater imbalance between input power and radiative losses results which drives the temperature up even faster. This radiative instability continues until the plasma can cool either by Bremsstrahlung or, more likely, by thermal conduction to cooler regions. (We note that radiative instability in the absence of heat input is also the basic reason why cool, dense prominences form in the upper outer atmosphere: as plasma cools, it radiates more efficiently and cools even more rapidly [e.g., Tandberg-Hanssen (1974)].) The chromospheric plasma which is explosively heated by the above process is violently thrust up the legs of the loop and into its coronal parts.

The observational evidence for chromospheric evaporation comprises Doppler-shifted spectral lines emitted from the upward-moving flare plasma when viewed on the solar disk. Doschek *et al.* (1980), Feldman *et al.* (1980), and Antonucci *et al.* (1982, 1984) have indeed observed disk events in soft X-ray lines from Ca XIX and Fe XXV where the lines exhibit a blue asymmetry and have broad profiles. The blue shift may correspond to upward velocities of $300–400 \, \text{km s}^{-1}$ (Antonucci *et al.*, 1982, 1984), and the width substantially exceeds the thermal Doppler width. Spectra of limb flares do not show such blue-shifted lines (Antonucci and Dennis, 1983), but the excess broadening is still present. These authors interpret the observations in terms of a mainly radial, upward motion of the evaporating plasma, where the broadening is due to nonthermal turbulence.

Emslie and Alexander (1987) showed that an alternative interpretation also is possible, since the observed Ca XIX profiles, in both disk and limb events, are consistent with the hydrodynamics of atmospheres heated by thick-target electron bombardment; see Section 8.1.1. In their interpretation the broadening is due to the differential motion of plasma throughout the region of soft X-ray line emission in the flare loop; a ubiquitous feature of hydrodynamic models of the flaring atmosphere is that typical velocity dispersions are of the same magnitude as the mean flow velocity. Thus, although shifts are to be expected only in disk events, we still expect line broadening – due to differential line-of-sight velocities – to be present in all events.

8.1.4   *Thermal conduction*

As energy is suddenly released in the upper, coronal part of a flare loop, a temperature gradient is set up between this hot part and the cool chromosphere. Therefore, in addition to the energy transport by particles considered in the previous section, we also witness the effect of thermal conduction of energy.

For classical conduction [e.g., Spitzer (1962)], the conductive flux F_{cond} is proportional to the temperature gradient ∇T, and can be expressed in terms of the differential emission measure $Q(T)$, a quantity measurable through observation of spectral line intensities (see Section 2.6). Since

$$\nabla T = \left(\frac{p}{kT}\right)^2 \frac{S}{Q(T)}, \tag{8.20}$$

where p is the gas pressure and S is the flare, the conductive flow (erg s^{-1}) may then be written as

$$F_{cond} = KT^{5/2}S\nabla T = K\left(\frac{p}{k}\right)^2 \frac{S^2}{Q(T)} T^{1/2}. \tag{8.21}$$

For a class 2B flare studied by Withbroe (1978), equation (8.21) gives a total energy loss (flow × time), $E_{cond} = 1.4 \times 10^{31}$ erg for the gradual phase, if the area S was 1.3×10^{19} cm^2, if we use $Q(T)$ at a typical transition-region temperature of 10^5 K, and if the phase lasted about 1.2×10^4 s. This is comparable to the energy deposited by nonthermal electrons in an event of this size (Section 8.1.2).

In strongly heated flares, temperature gradients may be so steep that the classical conductive flux expression (8.21) gives a value of F_{cond} which is beyond the ability of the plasma particles to carry and/or the mean free paths of the particles are larger than the temperature gradient scale length, so that the electrons freely stream rather than diffuse. In such cases, the heat flux is given by the saturated heat flux F_{sat}, obtained by multiplying the thermal energy density of the plasma by the maximum streaming velocity v_s, viz.

$$F_{sat} = \left(\tfrac{3}{2}nkT\right)v_s = \tfrac{3}{2}nm_e v_e^2 v_s. \tag{8.22}$$

Numerical simulations (Manheimer and Klein, 1975) show that $v_s \approx v_e/6$, so

$$F_{sat} \approx \tfrac{1}{4}nm_e v_e^3. \tag{8.23}$$

In cases where F_{cond} exceeds F_{sat}, we must replace the former by the latter.

Another limitation to the heat flux comes from the collective plasma physics associated with a bulk streaming of electrons (see Section 4.5). Too high a velocity sets up ion-acoustic waves which enhance the collision frequency and reduce the heat flux that can diffuse outward. The plasma thus adjusts to a marginally stable (Manheimer, 1977) state, in which the growth rate of ion-acoustic waves is zero. This corresponds [cf. Smith and Lilliequist (1979)] to a so-called anomalously limited heat flux F_{an}, given by

$$F_{an} = \tfrac{3}{2}n_e m_e v_e^2 \left\{ \left[\frac{k(T_e + 3T_i)}{m_i}\right]^{1/2} + \left(\frac{m_i}{m_e}\right)^{1/2}\left(\frac{T_e}{T_i}\right)^{3/2}\left(\frac{k}{m_i}\right)^{1/2} \right.$$
$$\left. \times \left[(T_e + 3T_i)^{1/2} - T_i^{1/2}\right]\exp\left[-\tfrac{4}{5}(T_e/2T_i + \tfrac{3}{2})\right] \right\}. \tag{8.24}$$

For $T_e/T_i \gtrsim 8$, this represents a more stringent limitation than the saturated flux (8.23).

In summary, if the mean free paths are sufficiently small compared to the temperature scale length $T_e/|\nabla T_e|$, and if $F_{cond} < \min(F_{sat}, F_{an})$, then the classical heat conduction formula (8.21) is a valid representation. Otherwise, the flux is reduced below this value; we refer the reader to Smith and Lilliequist (1979) for the appropriate results.

Heat flux limitation by both saturation and anomalous effects is especially important in the transition region, with its steep temperature gradients. Indeed, as pointed out by Shoub (1983), the temperature scale lengths are so short compared to the mean free paths of the high-energy electrons that carry the bulk of the heat flux $(F \sim \int f(v)v^3 \, dv)$ that even the concept of a local temperature is meaningless. Shoub's results are derived for quiet Sun transition-region models and are even more pertinent to flares because of the even steeper temperature gradients involved (which more than offset the decrease in mean free path associated with the higher densities appropriate to flare conditions). Care must therefore be taken to adequately model heat flow through the transition region in flares and in interpreting the radiation signatures for this plasma.

8.1.5   *Radiative transport*

In addition to particle beams and thermal conduction, radiation often furnishes a very efficient vehicle for energy transport out of a hot flare plasma. In analogy with equation (8.21) for conduction we could define a photon flux, F_{rad}. Instead we go directly to the radiative loss, E_{rad}, from the flare plasma and use the differential emission measure [equation (2.50)] to write

$$E_{rad} = \int n_e n_H \phi(T) \, dV = 0.8 \int Q(T) \, dT, \tag{8.25}$$

where $\phi(T)$ is the radiative loss function (Figure 8.3), and the factor 0.8 holds for the relationship of n_e and n_H in a fully ionized solar plasma. Withbroe (1978) used values for $\phi(T)$ given by Rosner *et al.* (1978) and evaluated $Q(T)$ for a class 2B flare to deduce a radiative loss of $E_{rad} = 2.2 \times 10^{31}$ erg.

This value is comparable to the thermal conduction losses that Withbroe found for the same flare (Section 8.1.4), and we are led to the conclusion that both radiative and conductive losses must be considered as important when studying the physics of the gradual phase for flare plasmas of temperatures greater than 10^5 K.

In general terms, we may say that the energy balance in flare loops during the gradual phase is dictated by the conditions that the energy conducted down from the hot, higher parts of the flare loop must be dissipated and lost by radiation (e.g., Lα, Hα, other lines), conduction to the cooler chromosphere, and – if these losses are not sufficient – by heating of the chromospheric plasma which then expands upward into the higher parts of the flare loops, i.e., undergoes chromospheric evaporation (see Section 8.1.3). The strong radiation in lines such as Lα and Hα, and in the X-ray and EUV continua (cf. Section 7.5) is also significant for energy transport into regions of the plasma, as well as a method of cooling it. Somov (1975), Machado (1978), and Hénoux and Rust (1980) have considered the role of soft X-ray absorption in heating the chromosphere – the soft X-ray continuum acts as a 'light bulb' which illuminates the atmosphere below where light is absorbed, mostly by photo-ionization of hydrogen and other elements (Somov, 1975). This process of so-called *radiation backwarming* can be a substantial source of chromospheric heating and is probably responsible for the fainter 'halo' seen in Hα around the more intense kernels, presumably produced by a process like electron bombardment. Since photons, unlike electrons, are not required to follow magnetic field lines, their heating effects are visible over a wider area than that corresponding to the magnetic field lines on which the primary energy release occurs. Somov (1981) suggested that intense electron heating, coupled with a surrounding region of weaker soft X-ray heating, is responsible for the 'double trench' shape of the EUV emitting plasma inferred from center-to-limb studies of EUV flare intensities (Donnelly and Kane, 1978); see Section 7.5.1. Falciani *et al.* (1977) found a close relationship between the temporal behavior of Hα and soft X-rays in flares, indicating that soft X-ray heating probably accounts for part of the chromospheric flare response. On the other hand, Machado *et al.* (1978) have attempted, with negative results, to explain the strong heating found at temperature minimum levels in flares by means of soft X-ray irradiation. These levels correspond to column densities $N \approx 10^{23}$ cm^{-2} and are well below the depth of penetration for deka-keV nonthermal electrons. However, Machado *et al.* (1986) pointed out that irradiation by the λ 1548-Å line of C IV (formed in the newly heated high-density flare transition region; Section 8.1.3) can result in substantial photo-ionization of Si I to Si II at temperature minimum levels. The photoelectrons produced associate with ambient hydrogen atoms to produce H$^-$ ions and these, in turn, being the dominant source of atmospheric opacity at such levels, absorb large quantities of photospheric radiation, sufficient to explain the temperature enhancements inferred from the line profile of the Fraunhofer K-line of Ca II. Thus, not only is radiation a direct source of atmospheric heating, but it can also be an indirect means of accomplishing significant transport of energy throughout the flaring atmosphere (Machado *et al.*, 1988c). Due to the rather complicated nature of optically thick radiative transport theory, when applied to a plasma containing numerous species of ions with their associated thickness of lines and continua, our understanding of this area of flare physics is currently at a fairly basic level. Improved numerical techniques for dealing with optically thick radiative transport in time-dependent flare heating calculations [cf. McClymont and Canfield (1986)] point the way to substantial progress in this area.

8.1.6   *Mass motions*

Craig and McClymont (1976) pointed out that the strong temperature gradients established in flare atmospheres lead to large pressure gradients and, consequently, to significant mass motions. These motions represent an additional part of the flare energy budget which must be established by the primary heating source (such as nonthermal electron bombardment) and which subsequently deposit their energy into the thermal plasma through viscous dissipation. The changes in density which these motions produce also significantly affect the radiative loss profile throughout the atmosphere, as well as the spatial distribution of the heating function. The latter effect can be understood when we remember that the electron heating functions of Section 8.1.1 are defined in terms of the column density $N = \int n(s)\,\mathrm{d}s$; whence we see that the spatial form of the heating $I_{\mathrm{B}}(s)$ changes with changes in the $n(s)$ profile.

Numerous hydrodynamic simulations of the response of the pre-flare atmosphere to flare energy input have been carried out in recent years (McClymont and Canfield, 1983; Pallavicini *et al.*, 1983; Nagai and Emslie, 1984; Emslie and Nagai, 1985; Fisher *et al.*, 1985a, b) and we refer the reader to these papers for details.

8.2   Flare-loop structure and evolution

As a result of the response to the energy release, the flare plasma will consist of regions at different temperatures distributed along the flare loop. Different spectroscopic diagnostics are available to determine these temperatures as well as the density and velocity of the plasma. It is practical to distinguish between the high-temperature (coronal) flare plasma, with $T_e \geqslant 10^6$ K, and the low-temperature plasma, with $T_e < 10^6$ K. At all but the lowest temperatures the distribution of ionization states (responsible for the spectral lines used in the diagnostics) is mainly determined by the balance between collisional ionization and radiative and dielectronic recombination [see equation (2.16) and Section 2.2.5].

8.2.1 *The high-temperature flare*

The high-temperature plasma found in flare loops in the gradual phase radiates strongly in the soft X-ray region, where emission lines from highly ionized Ca and Fe dominates the 1–20-Å spectrum. The theoretical basis for using these spectral lines as temperature and density diagnostics has been worked out by Bhalla *et al.* (1975) and Bely-Dubau *et al.* (1979a, b), and the adopted nomenclature for the line spectra was introduced by Gabriel (1972). The unique diagnostic potential rests on the presence of a group of dielectronic satellite lines situated on the long wavelength side of helium-like resonance lines. We shall study in some detail the examples furnished by the resonance lines of Ca XIX and Fe XXV at 3.17 Å and 1.85 Å, respectively, which have been particularly well observed during the 1980 solar maximum, but first we consider some general properties of the spectrum at these wavelengths.

The emission comes from $n = 2 \rightarrow 1$ transitions in highly ionized ions of Ni, Fe, Ca, Ar, S, Si, Mg, Ne, and O, and the important lines occur mainly in one of the following types of ions (see Section 2).

(i) H-like ions, giving lines of Lyman series;
(ii) He-like ions, giving

resonance lines, designated	$w: 1s^2\,{}^1S_0 - 1s2p\,{}^1P_1^0,$
forbidden lines, designated	$z: 1s^2\,{}^1S_0 - 1s2s\,{}^3S_1,$
intercombination lines, designated	$x: 1s^2\,{}^1S_0 - 1s2p\,{}^3P_2^0,$
	or $y: 1s^2\,{}^1S_0 - 1s2p\,{}^3P_1^0;$

(iii) Ne-like ions, with transitions, e.g., $1s^2 2s^2 2p^6 - 1s^2 2s^2 2p^5 3l$;
(iv) Ions in which dielectronic transitions produce satellite lines;
(v) Ions between Fe XVII and Fe XXV undergoing $2p-1s$ transitions, e.g., $1s2s^2 2p^{q+1} - 1s^2 2s^2 2p^q$, resulting in lines in the 1.85–1.91 Å range;
(vi) Fe II ions and other near neutral ion species responsible for the inner shell transitions that lead to the $K\alpha_1$ and $K\alpha_2$ lines at 1.936 Å and 1.940 Å respectively (Section 7.6.1).

To use the spectroscopic flare observations and deduce parameters like temperature and density, practical approximations to the formulae of Sections 2.1 and 2.2 must be considered.

The relaxation of the atomic level population (labeled n_j for an upper level, n_i for a lower level of a transition) has a time constant that is very short compared to other reactions during the gradual phase. We therefore assume a steady state solution for the level populations (Bely-Dubau and Gabriel, 1984). Further, we consider optically thin emission lines, since the densities are fairly low in coronal loops. The intensity in a line is then given by equations (2.22) and (2.23).

The ionization equilibrium relaxes with a time constant longer than the time constant for the atomic level populations because, while the latter is related to the spontaneous radiative decay [equation (2.8)], the former is determined by slower recombination rates. The time-dependent expression for ionization equilibrium is then

$$\frac{dn^q}{dt} = n_{el}n^{q-1}P_{ion}^{q-1} + n_{el}n^{q+1}P_{rec}^{q+1}$$

$$- n_{el}n^q P_{ion}^q - n_{el}n^q P_{rec}^q, \tag{8.26}$$

where n^q denotes the ground level density of the emitting ion in the qth ionization stage, and n_{el} is the elemental abundance. The rates for ionization and recombination are P_{ion} and P_{rec} respectively. Note that for simplicity all ion populations, n^q, n^{q-1}, n^{q+1}, are assumed to be in the ground levels. Further, again for simplicity it is often assumed that a steady state prevails, and in the gradual phase this is a good approximation. Equation (8.26) then reduces to

$$\frac{n^{q+1}}{n^q} = \frac{P_{ion}^q}{P_{rec}^{q+1}}. \tag{8.27}$$

Under the coronal conditions that we consider the ions responsible for the observed emissions are mainly excited by collisions, i.e., $P_{ij} \approx C_{ij}$ in equation (2.7) or

$$\varepsilon = \frac{n^q}{n_{el}} \frac{n_{el}}{n_H} \frac{n_H}{n_e} n_e^2 C_{ij} h\nu. \tag{8.28}$$

In the solar atmosphere $n_H/n_e \approx 0.8$, and the observed intensity at the Earth's distance D from the Sun is then given in terms of the emissivity by

$$I = \frac{1}{4\pi D^2} \int \varepsilon \, dV \quad \text{erg cm}^{-2}\text{s}^{-1} \tag{8.29}$$

where one integrates over the emitting volume V [see Gabriel and Jordan (1972)].

8.2.1.1 Temperature determinations

During the 1980 solar maximum, extensive use was made of observations of the dielectronic satellite lines mentioned above to deduce the electron temperature of the flare plasma during the gradual phase.

Dielectronic transitions are transitions from doubly excited levels above the first ionization limit, for example (see Section 2.2.5). If the line originates at a level excited by dielectronic recombination, we designate it j; if the level is excited by an inner shell excitation (Section 2.2.4), we label it q or β. The strong satellite lines originate from levels with $n = 2$ for the perturbing electron. These are the lines we find in the spectra on the long wavelength side of the resonance line w. Other satellite lines belonging to $n = 3$ or higher lie closer to w and blend too much to be useful in most diagnostics.

Gabriel (1972) showed that the relative intensities of the satellite lines scale as Z^4 and are scarcely detectable in oxygen and neon. They are stronger in magnesium and silicon, but only in calcium and iron have they been used as an excellent temperature diagnostic in conjunction with the corresponding resonance line, w. In particular, we use the line intensity ratios k/w for Ca XIX around 3.18 Å and j/w for Fe XXV at 1.85 Å. Let us first look at the case where the satellite line is formed by dielectronic recombination. Then both the resonance line $I(res)$ and the satellite line $I(sat)$ are formed by electron interaction with the same ion stage, and the ion population cancels. We obtain, with Bhalla *et al.* (1975),

$$\frac{I(sat)}{I(res)} = F_1(T_e)F_2(sat), \tag{8.30}$$

where

$$F_2(sat) = g(sat)\frac{A(R)A(auto)}{\sum A(R) + \sum A(auto)}$$

only depends on the atomic parameters of the satellite line involved. The auto-ionizing rate is $A(auto)$; see equation (2.29). The factor $F_1(T_e)$, which depends on temperature and atomic constants but not on the satellite line, cannot be expressed analytically since it includes the rate coefficient $C(res)$ for excitation of the resonance line. It can be expressed by the approximate formula

$$F_1(T_e) = \text{const}\,\frac{1}{T_e}, \tag{8.31}$$

whence (8.30) takes the form

$$\frac{I(sat)}{I(res)} = \text{const}\,\frac{1}{T_e}. \tag{8.32}$$

On this functional form rests the importance of the use of the satellite lines as temperature diagnostics – when dielectronic recombination is the predominant excitation mechanism.

Second, we consider the case when inner shell excitation of the satellite line is important. In the line ratio $I(sat)/I(res)$, we then have to take into account the relative ion stage populations involved, n^{q-1}/n^q. With Bely-Dubau and Gabriel (1984), we can write

$$\frac{I(sat)}{I(res)} = \frac{n^{q-1}}{n^q} \frac{C(sat)}{C(res)} \frac{A(R)}{\sum A(R) + \sum A(auto)}, \tag{8.33}$$

where $C(sat)$ is the rate of inner shell excitations [see Dubau *et al.* (1981), Doschek *et al.* (1981) and Bely-Dubau *et al.* (1982a, b) for details].

The spectra of the flare of 9 May 1980 have been analyzed using the above theory to deduce the electron temperature. Figure 8.4 shows the Ca XIX resonance line *w* with its satellite lines as observed from the Solar Maximum Mission. To calculate a theoretical best fit the relative ion stage populations and effective Doppler temperature for the line widths as well as the electron temperature were adjusted. As is usual, T_e was determined from equation (8.32). The resulting best fit is shown as a smooth curve in Figure 8.4 with the final parameters chosen. We see that an electron temperature of $T_e = 1.6 \times 10^7$ K was deduced for the Ca XIX plasma. Note, however, that

Fig. 8.4. Ca XIX spectrum observed during the gradual phase of the flare on 9 May 1980 and compared with the theoretical spectrum (smooth curve) for an electron temperature $T_e = 1.58 \times 10^7$ K, an equivalent ion temperature $T_i = 1.70 \times 10^7$ K, a Li to He-like ion abundance ratio Li/He = 0.19 and a H to He-like ion abundance ratio H/He = 0.16 (after Antonucci *et al.*, 1982).

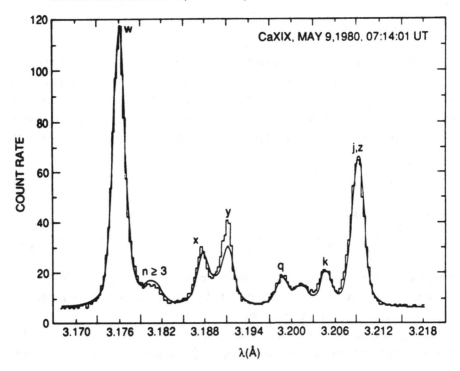

the temperature is a suitably weighted mean value (Emslie, 1986); what is really required, and not yet available, is an observational determination of the differential emission measure function $Q(T)$ (Section 2.6) at temperatures corresponding to the flaring corona. This requires simultaneous observation of lines formed over a wide temperature range and is within the scope of the current instrumentation.

For somewhat lower temperatures, i.e., several times 10^5 K to several times 10^6 K, other line-intensity ratios are used for their determination. A commonly used diagnostic for He-like ions is the intensity ratio

$$G = [I(1s^2\,^1S-1s2s\,^3S) + I(1s^2\,^1S-1s2p\,^3P)]/I(1s^2\,^1S-1s2p\,^1P).$$

Between $T_e = 6 \times 10^5$ K and 2×10^6 K, the quantity G varies, however, only by a factor of 1.8. Recent calculations (Tayal and Kingston, 1984) of electron excitation of the O VII ion have been used by Keenan *et al.* (1985) to develop new diagnostics in this temperature range, where it turns out that the following intensity ratios R_1 and R_2 vary by factors 3.3 and 4.9 respectively. Here $R_1 = I(1s^2\,^1S-1s3p\,^1P)/I(1s^2\,^1S-1s2p\,^1P)$ and $R_2 = I(1s^2\,^1S-1s4p\,^1P)/I(1s^2\,^1S-1s2p\,^1P)$.

8.2.1.2 *Density of the high-temperature plasma*

Metastable levels exist in many ions, and we have seen in Section 2.2.2 that certain forbidden lines arise when an electron jumps from such a level. Normally these lines are quite weak, but under certain conditions the population of a metastable level may rise significantly compared with the population of the ground level, and strong forbidden lines occur. In the tenuous corona, the metastable level population is proportional to the electron density, and the ratio of a forbidden line and an allowed line [which is constrained by the selection rules, equation (2.25)] can then be used as a density diagnostic. An important example is furnished by the metastable level $1s2s\,^3S_1$ in He-like ions, which under coronal conditions can decay via the forbidden transition z to $1s^2\,^1S_0$. In Fe XXV this line is found at 1.868 Å and in Ca XIX at 3.21 Å (see Figure 8.4). As the density increases the population of the $1s2s\,^3S_1$ level increases and the level $1s2p\,^3P$ will be populated from $1s2s\,^3S_1$. This again will lead to emission of the intercombination lines y: $1s^2\,^1S_0-1s2p\,^3P_1$ and x: $1s^2\,^1S_0-1s2p\,^3P_2$. In other words, as the electron density increases, the intensity of the z line is progressively transferred to the lines x and y. The ratios z/y and z/x therefore are sensitive to the electron density over a certain density range determined by the competition between the transitions from the $n_j \equiv 1s2s\,^3S_1$ level, i.e., a balance between collision to the level, e.g., $n_k \equiv 1s2p\,^3P_1$ or radiative decay to the $n_i \equiv 1s^2\,{}^1S_0$ level:

$$C_{jk}n_jn_e \approx R_{ji}n_j. \qquad (8.34)$$

Wolfson *et al.* (1983) used the z/y ratio during the gradual phase of the flare of 5 November 1980 and found that n_e in the Ne IX-emitting plasma decreased from about $10^{12}\,\mathrm{cm}^{-3}$ to $3 \times 10^{11}\,\mathrm{cm}^{-3}$ over a roughly 20-min period.

8.2.2   *The low-temperature flare*

As discussed in Section 1.2.2, the low-temperature flare, notably in Hα, has long been the primary evidence of the flare process. This was largely due to the lack of observations at other wavelengths, such as hard X-rays, which perhaps give a better indication of flare-energy release. Nevertheless, the low-temperature flare still possesses great diagnostic value, and not all aspects of it have been fully explained.

8.2.2.1   *The Hα flare*

The detailed morphology of Hα flares is a complex subject, and we refer the reader to the excellent book by Švestka (1976) for a thorough review. Here we simply summarize the salient points.

In Section 1.2.2 we broadly classified Hα flares into *compact* and *extended* events. Typical of the latter category is the two-ribbon flare, which we can now identify with the footpoints of an arcade of flaring loops. Possibly in response to reconnection following filament eruption (Section 9.1), higher and higher field lines develop regions of energy release, and the resulting pattern of Hα ribbons shows a tendency to spread outward, away from the magnetic neutral line. We identify compact Hα flares with small flaring loops, or sets of loops, whose individual bright footpoints (energized both by particle and radiation heating) are too close together to be individually resolved.

Using charge-coupled device (CCD) arrays, it has recently become a fairly routine matter to examine the behavior of the Hα line profile as the flare proceeds. The impulsive phase of a flare is characterized by broad Hα wings, which are interpreted as being due to collisional Stark broadening (Section 2.5.1) in the high-density regions of the flare chromosphere. Later in the event, these wings decay as the flare returns to its pre-flare state (Section 8.3.2). Canfield *et al.* (1984) modeled the effect of varying the coronal overpressure, the conductive flux, and the nonthermal electron flux on the shape of the Hα profile and concluded that the broad wings observed are strong evidence for thick-target electron bombardment. The time-dependent behavior of the Hα profile has been studied by Canfield and

Gayley (1987). They pointed out that the fastest variations, arising from rapid heating, are evident on timescales down to a tenth of a second and principally affect the line center; slower variations (of the order of a second or greater), due to both ionization imbalance and hydrodynamic effects, affect primarily the blue wing and red wing, respectively. Future observations of Hα profiles with high time resolution are therefore capable of investigating numerous facets of the transport of flare energy.

Lemaire *et al.* (1984) studied the behavior of other important low-temperature lines (e.g., Lα, Lβ, and Hε of H I; H and K of C II; and h and k of Mg II) during the development of a flare and found that the observations indicate a downward propagation of energy as the flare proceeds. As the heating propagates from the level where Hα is formed down into deeper layers of the chromosphere, the intensity of the Lα line first reaches maximum intensity, followed by the Lβ line, 20 s later, and then the Ca II and Hε lines exhibit their maximum intensity, another 20 s later. Studies of the Hε profile, at the lowest height, led Lemaire *et al.* to estimate that the electron density in the relaxing flare plasma was less than $8 \times 10^{13}\,\mathrm{cm^{-3}}$.

8.2.2.2 *The temperature minimum region*

The temperature minimum region on the quiet Sun occurs at a column depth $N \approx 10^{21}\,\mathrm{cm^{-2}}$, where the minimum temperature is ≈ 4200 K. Below this level lies the solar photosphere and interior; above it the chromosphere, transition region, and corona. During solar flares, as a result of the overall heating of the atmospheric layers by the flare energy input, both the temperature and the column depth of the temperature minimum increase to around 4400 K and $10^{23}\,\mathrm{cm^{-2}}$, respectively, as inferred from *empirical* modeling.

The process of empirical modeling has its basis not in the solution of the hydrodynamic equations describing the response of the solar atmosphere to a prescribed form of energy input, but, rather, in the determination of that structure which best reproduces observed flare features at a variety of wavelengths. The technique has been used with considerable success in modeling the quiet Sun (Vernazza *et al.*, 1976, 1981), active regions (Basri *et al.*, 1979), and flares (Machado and Linsky, 1975; Machado *et al.*, 1980). Despite problems of nonuniqueness in the fitting process [cf. Craig and Brown (1985); Emslie *et al.* (1981)], the method provides a reasonably reliable estimate of gross changes in atmospheric structure. Using this technique to model observations of the Fraunhofer K-line, Machado *et al.* (1978) inferred the above-mentioned temperature enhancements of several hundred degrees Kelvin at temperature minimum levels in flares, raising the actual minimum temperature by some 200 K and pushing it to a particle column density of $N = 10^{23} \, cm^{-2}$.

The corresponding level is significantly deeper than it is possible for the bulk of the high-energy electrons to penetrate. According to equation (4.76), it corresponds to the stopping distance for 1-MeV electrons. Further, neither bombardment by protons nor irradiation by soft X-rays or EUV radiation can deposit sufficient energy (Machado *et al.*, 1978; Emslie and Machado, 1979). The latter authors suggested that some form of local heating may be responsible, e.g., joule dissipation of steady currents in the relatively high-resistivity layers of the temperature minimum region. Following this suggestion, Emslie and Sturrock (1981) proposed that these currents were the result of Alfvén waves propagating from the energy-release site and were able in fact to account for the required energy deposition rate, but only for a narrow range of frequencies and amplitudes of the Alfvén waves and for a similarly narrow range of atmospheric magnetic field structures. Recently, Machado *et al.* (1986) pointed out that irradiation by the 1548 Å line of C IV produces photo-ionization of silicon atoms; these photoelectrons subsequently combine with hydrogen atoms to form H^- ions that can absorb sufficient photospheric radiation to account for the observed temperature enhancement.

8.2.2.3   *White-light flares*

The first recorded observations of a solar flare were obtained independently by Carrington (1859) and Hodgson (see Clerke, 1903). What these observers witnessed was an intense white-light flare, so named because the radiation level of the photospheric continuum is enhanced.

White-light flares (WLFs) have proved to be an extremely enigmatic phenomenon. They used to be considered very rare, but this now appears to be simply a question of sensitivity, since it is difficult to detect weak WLFs against the strong photospheric background (see Section 1.1). Kane *et al.* (1985) showed that bursts in white-light flare emission occur simultaneously with hard X-ray bursts, implying a rapid excitation. However, they are also observed to persist long into the event, and so they more properly belong in the gradual phase, according to our (somewhat arbitrary) division.

One of the most challenging puzzles of white-light flare emission is the nature and source of the continuum spectrum. Machado and Rust (1974) reported a preferential enhancement toward the blue, and suggested that ionization of hydrogen atoms, followed by Balmer recombination, was the responsible agent [see also Hudson (1972)]. Hiei (1982) suggested instead that the H^- ion is the source of the radiation. This would indicate a deep level of formation of the enhanced continuum and amplify the problems of energizing it impulsively. On the other hand, Neidig and Wiborg (1984) reported evidence of both Balmer and Paschen jumps in the spectrum of a white-light flare and suggested that the flare is an upper chromospheric phenomenon, produced by bound–free continuum recombination onto hydrogen [see also Donati-Falchi *et al.* (1985)]. Machado *et al.* (1988*c*) suggested that Lyman photo-ionization by strong EUV continuum fluxes (Section 7.5) may be responsible for creating the necessary photoelectrons for the recombination continua observed. We conclude, therefore, that at the present time the exact origin and cause of white-light flares remains largely controversial.

8.3   Relaxation to the pre-flare state

After the energy input has ceased, the flare atmosphere relaxes (radiatively, conductively, and hydrodynamically) to its initial pre-flare state. The relaxation of this phase, sometimes referred to as the *decay phase*, has been studied extensively and provides important insight into the nature of the post-flare plasma.

8.3.1   *Post-flare loops*

Post-flare loops are frequently observed to emit brightly in soft X-rays, EUV, and Hα. Their presence is a strong indication that the essential magnetic topology of the active region is not wholly destroyed by the dissipative processes of the flare – at least the potential component of the magnetic field appears to remain. In fact, Kiplinger *et al.* (1983) reported observations of successive multiple bursts in an active region. They interpreted the repetitive pattern of the successive bursts as indicating a re-flaring of the same ensemble of loops.

As mentioned in Section 1.3, post-flare loops are normally seen in absorption on the disk in Hα. However, at times, anomalously dense loops occur and show up in emission in the disk. Computations show that this behavior requires the electron density in the loops to be close to $10^{12} \, \text{cm}^{-3}$ (Švestka *et al.*, 1986), a value that has interesting implications for the cooling rate of these structures.

Post-flare loops have a thermal structure very similar to that of nonflaring active-region loops; the assorted 'loop scaling laws' (Rosner *et al.*, 1978) seem to hold in post-flare loops also (Machado and Emslie, 1979). Having a cooling rate in excess of the quiescent rate of energy supply, they cool down to quiescent temperatures, as discussed below.

8.3.2   *Relaxation models*

There are three timescales of interest to the relaxation of post-flare loops to the quiescent state, viz.

1. Radiative timescale

$$\tau_r = \frac{3nkT}{n^2\phi(T)},$$
(8.35)

2. Conductive timescale

$$\tau_c = \frac{3nkT}{KT^{7/2}/L^2},$$
(8.36)

3. Hydrodynamic relaxation timescale

$$\tau_h = \frac{L}{(kT/m_p)^{1/2}},$$
(8.37)

where $\phi(T)$ is the optically thin radiative loss function (Raymond *et al.*, 1977; cf. Section 8.1.3). Depending on the parameters of the flare loop, either τ_r or τ_c will be dominant; the hydrodynamic timescale serves merely as a time-scale for redistribution of energy and for variation of the density, and therefore for variations in τ_r and τ_c. The overall release of energy from the flare loop occurs on a timescale

$$\tau \approx (\tau_r^{-1} + \tau_c^{-1})^{-1}.$$
(8.38)

Energy 'lost' through conduction is ultimately radiated away, by the dense chromospheric layers, with a high degree of efficiency. Various models of the decay phase of the flare have been investigated with the role of radiative, conductive, and hydrodynamic processes taken into consideration in varying degrees (Moore *et al.*, 1980). A notable point regarding τ_c is the possible role of a converging magnetic field, which acts to 'choke off' the escaping conductive flux, and so increase τ_c (Antiochos and Sturrock, 1976).

Models that neglect radiative losses fail to produce the required variation of the differential emission-measure function $Q(T)$ throughout the flare (Underwood *et al.*, 1978; Canfield, 1979; Machado and Emslie, 1979). Whether or not hydrodynamic effects are important is less clear (Machado and Emslie, 1979). During periods of rapid cooling, the radiative instability associated with the increase in radiative loss rate with decreasing temperature (Figure 8.3), may lead to the formation of cool dense regions in the post-flare loop (Antiochos, 1980).

9

Coronal mass ejections

All quit their sphere, and rush into the skies

An Essay on Man, A. Pope (1688–1744)

Many flares are observed to be associated with the ejection of matter into the outer corona and sometimes beyond. These *coronal mass ejections* (CMEs) discovered during the Skylab era (Hildner, 1977) may carry a significant amount of energy and are often an important part of the flare phenomenon. Other CMEs cannot be associated with observed flares, but are related to eruptive quiescent prominences.

The term *coronal transient* refers to sudden changes in coronal structures as seen in coronagraph observations. Mass ejections cause some of these changes, but other transients are produced by wave actions. For good reviews of CMEs the reader is referred to Munro *et al.* (1979), Dryer (1982), Wagner (1984), and Hundhausen *et al.* (1984). Simon (1987) studied different physical processes that may lead to mass ejections.

9.1   Flares and prominences

Considerable uncertainty – even controversy – exists when one considers the role of prominences in the flare phenomenon, or the relationship between prominences and flares. It is known that certain types of prominences occur after flares, e.g., loop prominences, and they are to be considered part of the flare display, called post-flare loops. It may be a matter of opinion – or semantics – whether we refer to these loops as prominences or flares. Even though spectroscopy shows that the optical spectrum from post-flare prominences differs in details from the spectra from other parts of the flare emission [e.g., emission patches in the high chromosphere (Tandberg-Hanssen, 1963)], Zirin and Tandberg-Hanssen (1960) showed that, in general, low-lying limb flares and loop prominences exhibit very similar optical spectra. These are the loops that bridge the ribbons in the two-ribbon flare display (see Section 1.4).

Of great interest is the role of active-region prominences – active filaments – that are suddenly expelled. Particularly violent expulsions are called flare *sprays* (Warwick, 1957; Tandberg-Hanssen *et al.*, 1980), a name some authors reserve for those moving out with velocities reaching the escape velocity from the Sun.

The behavior of the high-velocity prominence plasma is an important key to our understanding of flare triggering. We have seen in Section 7.1.6 that the force responsible for the expulsion is due to a rearrangement of the magnetic field structure, including the prominence-supporting magnetic field. After the expulsion of the filament, the chromospheric response is dominated by the characteristic two-ribbon flare display. Only the most violent active-region filament eruptions lead to CMEs in which the velocity of the erupting plasma exceeds the escape velocity. In the more common filament eruptions the velocities are much smaller, but the ejected matter still may be associated with a well-observable coronal mass ejection.

A third relationship exists between quiescent prominences that are suddenly expelled out into the corona, the classical 'disparition brusque' phenomenon in the French nomenclature, and the ensuing flare brightening at chromospheric levels. This brightening is again a two-ribbon flare, where the large ribbons often are discontinuous and patchy (Tang, 1985).

With this background the question arises of how one should interpret the reports of observed prominence eruptions without any accompanying flare display. When the eruptive prominence is observed against the solar disk, the associated chromospheric flare emission is normally present. It is therefore likely that in those cases where we observe the eruptive prominence above the solar limb, and when no flare is reported, the flare emission is observed or indistinguishable from low-level activity at the limb. In all cases one expects some – generally most – of the erupted mass to return to the chromosphere. Hence, after a prominence eruption we will observe chromospheric flaring, either due to impact of electrons accelerated in flare loops associated with the prominence eruption, as discussed in Section 8.1, or to heating caused by conduction from above, or maybe due to the – somewhat milder – heating caused by the infalling prominence material, i.e., the infall–impact process discussed by Hyder (1968).

These prominence-induced flarings were first described by Waldmeier (1938*a*) and discussed by Martres (1956) and Becker (1957). According to Bruzek (1951, 1957) the brightening of the facular structure may be described as strings of bright mottles parallel to the disappearing filament; see also Tang (1985).

In cases like these, it has been argued that we are witnessing the formation of flaring regions at the expense of energy stored in the filament (Kleczek, 1963; Sturrock and Coppi, 1966; Hyder, 1967*a,b*, 1968). Hyder developed a model in which modifications of the magnetic field configuration lead to a falling of the prominence material down into the chromosphere which thereby becomes heated and shows up as a brightening. This infall–impact mechanism for solar flares has been developed further by Nakagawa and Hyder (1969) and describes how the chromosphere responds to the shock waves that develop in front of the falling material. It is, however, very unlikely (cf. Section 6.2) that the resulting heating can compete favorably with heating due to conduction and to impact of accelerated electrons.

To the extent that the infall–impact mechanism is responsible, we conclude that the basic cause of these prominence-induced flarings is the disruption or realignment of the large magnetic field structures that support the prominences and that become unstable and cause their eruption.

For the flares discussed in Chapter 7, the disruption of the magnetic field is associated with reconnection and annihilation that are the causes of the complex and energetic flare phenomenon, characteristic of particle acceleration. It is uncertain to what extent reconnection plays a role during the realignment of the field that seems to cause the milder eruptions which lead to the infall–impact flarings. We are probably witnessing a class of disruptions of varying degree of severity. At one end of the scale we have fairly benign realignments, causing only smaller eruptions, not very important flares, no particle acceleration, and hence no hard X-ray or microwave emission. At the other extreme the disruption of the field is so severe that important annihilation takes place leading to significant particle acceleration with the ensuing hard X-ray, γ-ray, and microwave emissions.

9.2   Characteristics of solar mass ejections

During years of high solar activity CMEs are a common phenomenon. In 1980 SMM observations indicated an average 0.9 CME per day (Hundhausen *et al.*, 1984), while *Solwind* observations led to an estimate of 1.8 CME per day for all CMEs, and 0.9 per day for major CMEs (Howard *et al.*, 1985). During the minimum activity year 1984 both SMM and *Solwind* observations were consistent with a frequency of occurrence for CMEs $\frac{1}{6}$–$\frac{1}{5}$ of the frequency in the maximum year of 1980.

As we have discussed previously (Section 7.1.7), neither the flare nor the prominence 'causes' the CME; rather all three phenomena are due to disturbances of the global magnetic field configuration in and around the site of the flare or prominence; see also Webb and Hundhausen (1987).

Figure 9.1 shows a CME as observed by the Coronagraph/Polarimeter on the SMM. We notice a three-part structure: out in the corona, mass – visible by electron scattering – is being ejected in the form of an outgoing front, which is followed by a 'void', a dark area of the corona. Finally inside, and lower than the void, cooler material, observable in Hα, is moving out with velocity normally smaller than the above-mentioned front; this is the eruptive prominence – or flare material.

On many coronagraph images of mass ejections one can distinguish an outer region, which leads the loop or blob-like ejection. This region, where the density is slightly enhanced over the pre-ejection corona, is referred to as the *forerunner* (Jackson and Hildner, 1978). The mass involved in the forerunner as compared to the following mass ejections is not negligible; Jackson and Hildner estimated that from 10 % to perhaps as much as 25 % of the total mass resides in the forerunner. The outer part of the flare forerunner travels out through the corona roughly 1 $R_\odot$ in front of the main mass ejection.

9.2.1   *Velocity*
In Figure 9.2 the histograms show the distribution of speeds for flare-related and prominence-related CMEs as observed during the Skylab

Fig. 9.1. Coronal mass ejection observed on 14 April 1980. See text for details (courtesy A. Hundhausen and High Altitude Observatory, National Center for Atmospheric Research).

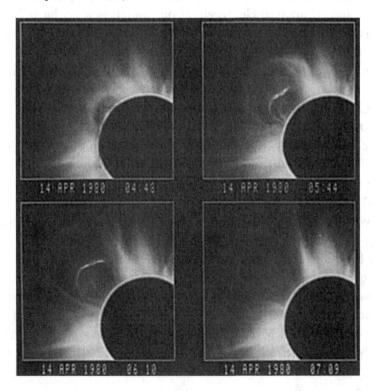

period (Gosling *et al.*, 1976). Since then numerous similar studies have been made that in essence confirm the earlier findings, even though there are differences related to the level of solar activity, the heliographic latitude of the mass ejections, and their speed (Hundhausen, 1987).

Fig. 9.2. Histograms of all measured speeds of the leading edges of coronal mass ejections (top panel) and similar histograms for CMEs associated with eruptive prominences, with flares, and with Type II and Type IV radio bursts (lower panels). Cross-hatching indicates lower limit estimates, and arrows give average values (from Gosling *et al.*, 1976).

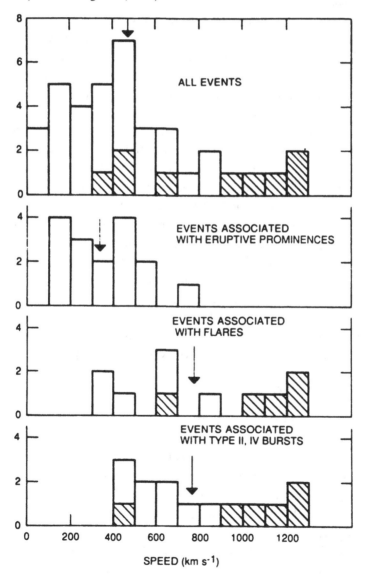

We see from Figure 9.2 that the outward speed of the ejections range from values less than 100 km s^{-1} to values in excess of 1200 km s^{-1}, and larger speeds also have been observed. The flare-related events tend to move faster than those events where no flare is observed but when prominence eruption takes place. It is worth noting that the flare-related events also have associated with them prominence eruptions, which is to be expected since we associate eruptive prominences with the basic flare display (see Section 7.1.6).

9.2.2   *Mass and energy*

With mass ejections tunneling through the corona at velocities often of several hundred kilometers per second, we witness displays of large amounts of kinetic energy. Estimates of the mass involved and of the corresponding kinetic energy have been made by a number of authors. The mass in question is the excess mass observed in the field of view of a coronagraph. One assumes that the corona far from the ejection does not change significantly during the time the CME is observed, so that a known background mass can be subtracted from the estimated mass of the observed ejection. The material in the corona and in the ejection is visible in coronagraph observations due to Thomson scattering of photospheric radiation by free electrons (van de Hulst, 1950); see Section 5.1. In practice one subtracts, point by point in a coronagraph picture, the pre-ejection brightness from the brightness of the ejection. Using equations (5.3) and (5.6) we can then convert from brightness to electron density; for details see Hildner *et al.* (1975), who constructed contours of differential mass density $m(\mathbf{r})$ by ascribing a mass to each electron, i.e., $2 \times 10^{-24} \text{ g}$ for coronal material consisting of 90% H and 10% He, fully ionized. From the differential mass density $m(\mathbf{r})$, one obtains the total excess mass M of the ejection by integrating $m(\mathbf{r})$ over the coronagraph's field of view, i.e.,

$$M = \int_0^{2\pi} \int_{R_{\min}}^{R_{\max}} m(\mathbf{r}) \, da, \tag{9.1}$$

where the limits $R_{\min}$ and $R_{\max}$ denote the distances, from Sun center, to the lower and upper edges, respectively, of the coronagraph's field of view.

Howard *et al.* (1985) studied nearly 1000 CMEs observed in the period 1979–81 and found values for the masses ranging from 10^{14} to more than 10^{16} g. The majority of CMEs showed masses of a few times 10^{15} g; Howard *et al.* adopted $M = 4.1 \times 10^{15}$ g as an average value. To find the kinetic energies, $E_k = \frac{1}{2} M v^2$, associated with these massive ejections, we need a determination of the velocity. Howard *et al.* (1985) measured the velocity of the leading edge of the ejections and found an average value of $v = 470 \, \text{km s}^{-1}$, giving for the 'average' CME a value

$$E_k = 3.5 \times 10^{30} \quad \text{erg}.$$

This is a substantial fraction of the energy of a flare and shows that the ejection of matter often plays an important role in the overall energetics of the flare phenomenon.

9.2.3 *Radio signatures of coronal mass ejections*

Several types of meterwave radio bursts are often observed in conjunction with CMEs. Since the radio bursts can furnish important information on the physics involved, we shall briefly discuss the association of CMEs with such bursts. It is outside the scope of this book to treat the nature of solar meter radio bursts in detail; the reader is referred to excellent overviews by Steinberg and Lequeux (1963), Wild *et al.* (1963), Kundu (1982), Dulk (1985), and Pick (1986). Briefly speaking, Wild and McCready (1950) classified three different components in the solar radio emission according to their spectral properties, and called them Type I, Type II, and Type III. Type I and III bursts often are part of radio noise storms [see, e.g., Boischot (1958); Avignon and Pick-Gutmann (1959); Kundu (1965); Elgaröy (1977)], and high-resolution spectroscopy shows the often very complicated structure of Type I bursts (Elgaröy, 1961). U-bursts were first recognized by Maxwell and Swarup (1958) and constitute a subset of Type III bursts (see Section 7.6.2). Type IV bursts, with their broad (continuum) spectra, were first classified by Boischot (1957). They may come from a stationary source or from a source moving out through the corona. In the latter case the burst is referred to as a moving Type IV, or Type IV M burst. Finally a Type V continuum has also been classified (Wild *et al.*, 1959). It normally follows a Type III burst (Robinson, 1977); see also Weiss and Stewart (1965), Kundu *et al.* (1970), and Stewart (1972).

We now look at the association of Types I, II, III, IV, and IV M bursts with CMEs to see what insight we thereby can gain into the physics involved.

9.2.3.1 *Type II and IV radio bursts*

Type II radio bursts are due to plasma emission from electrons accelerated in shocks. Therefore, when we observe these limited frequency radio bursts from CMEs, we have a clear indication that the instability, responsible for the ejection of plasma, also produces shock waves in the corona.

On the other hand, Type IV radio bursts, broad frequency continuous emission that normally follow Type II bursts, may be accounted for by two different emission mechanisms, viz. plasma emission and gyrosynchrotron emission. Sometimes, observations of coronal transients at, e.g. 80 MHz and 40 MHz, show radio sources characteristic of harmonic plasma emission (Gary *et al.*, 1984); i.e., the 80-MHz source and the 40-MHz source are located in blobs in the corona whose densities are appropriate for the emission of plasma waves. According to plasma-wave theory (Section 3.4.1) the emission takes place near the plasma frequency [equation (3.75)]:

$$\omega_{pe}^2 = \frac{4\pi n_e e^2}{m_e}.$$

Since the density n_e decreases with height, h, in the corona, the 80-MHz radiation will come from a lower region than the 40-MHz emission. Therefore, having a model for the variation of n_e with h, one can map the emission at different frequencies in the corona. Figure 9.3 shows the location of the white-light transient in relation to the derived heights of the blobs responsible for the radio emissions. For another well-observed example, see Stewart (1984).

At other times it is possible that the magnetic field could have an intensity of 2–3 G, in which case emission from sufficiently energetic electrons may account for the radio emission as gyrosynchrotron radiation.

The moving Type IV bursts arise when the plasma blob, responsible for the emission of the burst, moves out through the corona as part of the overall mass ejection [see, e.g., Stewart *et al.* (1982); Gergely *et al.* (1984)]. If the density of the blob is high enough, the emission mechanism may again be plasma radiation (Duncan, 1981; Trottet *et al.*, 1981; Gergely *et al.*, 1984). In one case Stewart *et al.* (1982) found a blob whose density was 30 times higher than the background corona, amply sufficient for plasma radiation.

In some earlier interpretations the shock driven by the ejected mass was assumed to account for the Type II radio bursts. While there may be events where this is the situation, in other transients the speed of the CME is lower than that of the Type II burst source (Gergely, 1984; Gary *et al.*, 1984), and in other cases the Type II burst is located below the outward-moving top of the CME. Wagner and MacQueen (1983) inferred from this last type of event that the CME starts to rise first, and somewhat later the shock wave is generated that is responsible for the Type II emission. This shock wave may be generated when energy is released, e.g., in the impulsive phase of a flare. The Type II emission will then be generated as the shock on its way interacts with denser regions (generally the legs) of the CME. An example of this is shown in Figure 9.4 where the diagrams show the CME loop-shaped structures at two times during the ejection and compare them with the location of the 160- and 80-MHz Type II burst positions. It is clear from the diagram that the Type II sources are intimately linked to the CME structure.

9.2.3.2 *Type I noise storms*

Like Type IV bursts, Type I bursts are also due to continuum (broadband) emission. They are assumed to be due to coherent plasma radiation at frequencies in the range of the electron plasma frequency [equation (3.75)]. However, Type I emission differs from Type IV emission in that it seems independent of the CME. It exists before the ejection takes

Fig. 9.3. Comparison of radio source positions with visible light features in the 29 June 1980 event. The 80-MHz Type IV source (circled numbers 1–5) is always associated with the densest part of the mass ejection (after Gary *et al.*, 1985).

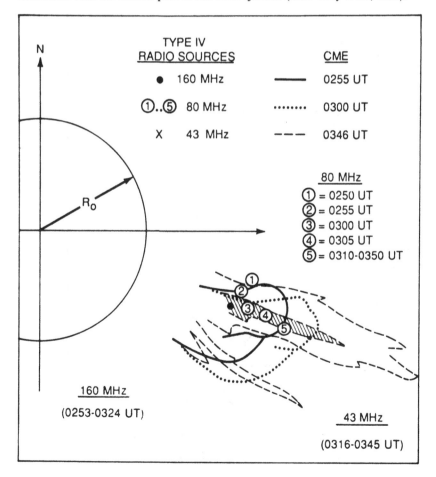

place and continues often long after the optical transient is over. It is therefore indicative of an ongoing activity above an active region that eventually gives birth to a CME. In the well-observed CME of 30 March 1980 (Lantos *et al.*, 1981; Lantos, 1985), Type I emission was observed high in the corona, at the level where 169 MHz is produced, before the loop-shaped transient occurred. The plasma waves may have been excited by

Fig. 9.4. Schematic diagrams of the mass ejection at two times: (*a*) 02:44:10 UT and (*b*) 02:47:30 UT, 29 June 1980 and comparison with the 160- and 80-MHz Type II burst positions (after Gary *et al.*, 1984).

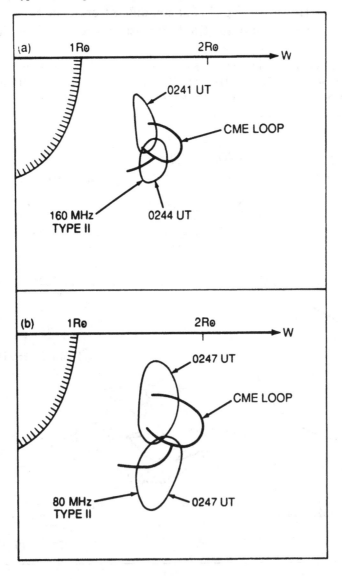

streams of electrons of energy 10–30 keV (Melrose, 1980; Benz and Wentzel, 1981), and the 169-MHz radiation, e.g., came from the level in the corona where $n_e = 3.5 \times 10^8 \, \text{cm}^{-3}$. The source of the noise storm was observed to move, and this motion therefore traces the $3.5 \times 10^8 \, \text{cm}^{-3}$ electron density level during the evolution of the coronal transient.

9.2.3.3 *Type III radio bursts*

We have discussed in Section 7.6.2 the flare-associated Type III bursts, who owe their existence to electron streams, accelerated in the impulsive phase of the flare. A connection between some CMEs and Type III bursts has also been made, since the bursts seem to be precursors of CMEs (Jackson *et al.*, 1980). While it is possible that electrons may be accelerated under pre-flare conditions, the connection with CMEs is difficult to establish due to the ubiquitous nature of Type III bursts (Gergely, 1986).

Leblanc *et al.* (1983) associated U-bursts with the initial phase of CMEs. In their picture the U-burst comes from the leading edge of the CME, which comprises an ejected (or rising) magnetic loop in which those energetic electrons travel that give rise to the radio emission. As the electrons travel along the loop where the density changes, the emitted frequency will change, as for other Type III bursts. However, in addition we will observe a turnover frequency, which gives the U-burst its name, as the electrons pass the apex of the loop. The velocity derived from the drift rate of the turnover frequency can then be identified as the velocity of the leading edge of the CME as it moves out in the corona. Leblanc *et al.* derived a mean velocity of 790 km s^{-1} for 35 events they studied, a value consistent with speeds found for CMEs.

9.2.4 *Magnetohydrodynamic disturbances*

Solar flares are at times seen to be the apparent source of wave-like phenomena propagating away from the flare site. When observed in Hα, this chromospheric part of the waves is referred to as a *Moreton wave* (Moreton, 1960; Athay and Moreton, 1961; Dodson-Prince and Hedeman, 1964; Moreton, 1964) which travels with a roughly constant velocity of 1000 km s^{-1}.

These waves are attributed to MHD fast-mode shocks generated in the impulsive phase of the flare (Anderson, 1966; Meyer, 1968; Uchida, 1968; Wild, 1969). The shock front expands quickly in the corona, and the skirt of the wave front sweeps across the chromosphere and is observed as a Moreton wave. The details of this 'sweeping skirt hypothesis' have been worked out by Uchida (1968, 1970) and Uchida *et al.* (1973). Uchida (1974*a*) also suggested that the same kind of MHD shock is responsible for the generation of Type II radio bursts in that a fairly weak MHD fast-mode shock may build up on encountering low-Alfvén velocity regions in the corona due to the focussing of the rays into such regions (Uchida, 1974b). In this picture Moreton waves and Type II burst shocks are two different facets of the same agent, viz. the flare-produced MHD, fast-mode shock front. For details the reader is referred to Uchida (1974a).

9.3   Origin of coronal mass ejections

Photographs of many CMEs reveal a loop-like structure that propagates out through the corona above an active prominence or a flare. Even though many other CMEs exist that do not show this simple, nearly axisymmetric picture, the loop structure is often considered to be the canonical shape of ejections, and most theoretical studies relate to such events. The question immediately arises whether the loop is a two-dimensional structure, or whether we in reality have to do with a three-dimensional 'bubble', that only looks like a loop due to the larger amount of matter observed near the edges as seen from the Earth. Depending on how one answers this question two schools of thought have developed as to the origin of CMEs.

9.3.1   *Compression wave model*

One early and reasonable interpretation considered the CME to be a bubble and explained it in terms of a compressive wave whose energy derives from the explosive energy release in the flare or prominence. Both gas dynamic shocks [e.g., Parker (1963); Dryer and Jones (1968); Burlaga (1971); Hundhausen (1972b); Dryer (1974)] and, in particular, MHD shocks (Nakagawa *et al.*, 1978; Nakagawa and Steinolfson, 1976; Steinolfson and Nakagawa, 1976; Steinolfson *et al.*, 1978, 1979; Wu *et al.*, 1978, 1982; Dryer and Maxwell, 1979; Dryer *et al.*, 1979) have been considered. As a result of these studies we have a good idea of how shock waves released from, e.g., flares will interact with the ambient coronal plasma and the solar wind (D'Uston *et al.*, 1981). When one compares the details of the observed shape of CMEs with the shapes predicted by the MHD or gas dynamic shock models, one often finds discrepancies. As a matter of fact, Sime and Hundhausen (1987) claim that those CMEs whose detailed shapes are well explained by the shock wave models are fairly rare, comprising maybe only a few percent of all CMEs. Nevertheless, our understanding of the response of the solar coronal plasma to explosive injection of energy is to a large extent due to the theoretical studies cited above. In addition, we now have data that clearly show the three-dimensional nature of one well-observed CME (Howard *et al.*, 1982), thereby confirming the reality of a 'bubble' model in some instances; see Figure 9.5.

Figure 9.6 shows the structure of a mass ejection as described by the shock wave model according to Dryer and Maxwell (1979), i.e., the response of the coronal plasma to the energy release associated with the triggering flare or eruptive prominence.

Most of the work on compression wave models has considered fast MHD shocks, and the prediction of their propagation in the corona leads to certain, well-defined structures in the CME. Some of these structural details are often in conflict with observations; e.g., the observed outward moving front is often flat, while the models predict a more or less spherical shape. Hundhausen *et al.* (1987) suggested that slow MHD shocks which form as the mass ejection sweeps out through the corona may resolve this conflict. They based their work on the observation that while most CME speeds, v,

Fig. 9.5. Alleged 'bubble-shaped' mass ejection observed on 17 November 1979, a so-called 'halo' coronal transient (courtesy E. O. Hulbert Center for Space Research, Naval Research Laboratory).

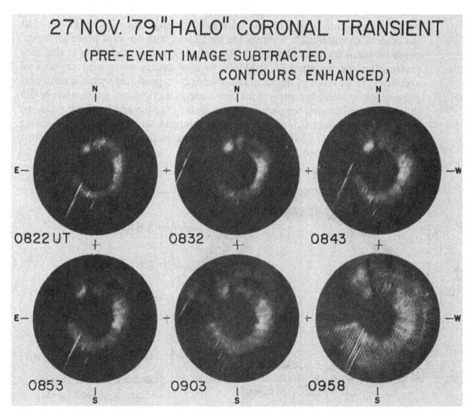

are greater than the sound speed in the corona, they are less than the estimated Alfvén speed, i.e., $c_s < v < V_A$. One can then show that the propagation of small-amplitude, slow-mode waves will lead to a flattened slow shock front, which Hundhausen *et al.* identified with the outward moving front of a CME.

9.3.2   *Two-dimensional flux tube models*

Observational data exist that make it difficult to avoid the conclusion that at least some CMEs are fairly planar loop structures (Trottet and MacQueen, 1980). In these models the CME is considered a rising loop-shaped magnetic flux tube, pushed out through the corona (Mouschovias and Poland, 1978; Anzer, 1978; Pneuman, 1980; Yeh and Dryer, 1981). While the models in Section 9.3.1 are derived by treating the MHD equations directly using numerical methods, the flux tube models are derived by using analytic approximations to account for the forces acting on

Fig. 9.6. Coronal density enhancement (compression) and depletion (rarefaction) computed by a two-dimensional MHD model (Dryer, 1982). Solid lines are excess density $(n - n_0)$ contours, as labeled; dotted lines are depletions. Values are based on an initially static, exponentially decreasing density with $n_0 = 3 \times 10^8 \, \text{cm}^{-3}$ at the base of an equilibrium atmosphere permeated with an ideal potential 'open' magnetic topology (after Dryer and Maxwell, 1979). The rarefaction at this time (6 min after the initiation of a 40-fold thermal energy pulse near the axis of symmetry) is the result of the simulated evaporation produced by the intense heat pulse which was used to represent the flare's energy conversion. A fast-mode MHD shock is formed around the perimeter, strongest at the top and weakening toward the sides; the contact surface is located at (approximately) the contour labeled $3.8 \times 10^8 \, \text{cm}^{-3}$. A slow MHD wave, which returns the flow to its initial state, forms deep within the rarefied zone.

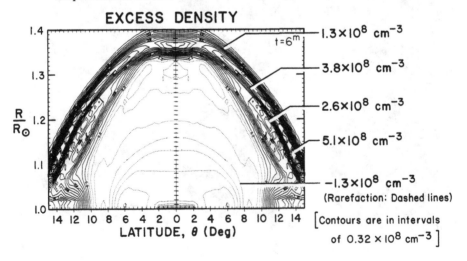

EXCESS DENSITY

the flux tube, thereby avoiding direct solutions of the complicated set of MHD equations.

One of the earliest of these often elegant models was developed by Mouschovias and Poland (1978), who considered a loop freely coasting out through the corona where gravity was exactly balanced by the local magnetic fields. Anzer (1978) studied the case where the driving force was due to the electromagnetic force a ring current exerts on itself. He showed that if a current loop can be produced fast enough, the subsequent motion of the loop is consistent with observations of loop-shaped CMEs. The model is based simply on the solution of the equation of motion, where only gravity, inertia, and electromagnetic forces are considered.

9.3.3   *The CME as the result of a global instability*

Both the compression wave models and the flux tube models shed light on several aspects of the nature of a coronal mass ejection, but as scenarios neither class of models seems capable of representing the complete phenomenon. Important improvements have been considered to take into account specific observational facts concerning the structure or dynamics of well-observed CMEs, but the complete set of events that constitutes a CME seems to require elements of both classes of models, as well as an interpretation in terms of a more global effect.

A CME is not caused by the blast wave generated in the initiation of the eruptive prominence/flare event. We have seen in Section 7.1.7 that the eruptive prominence and the flare are both caused by the overall rearrangement, and possible reconnection, in the global magnetic field surrounding the flare site. That reconnection is involved is made further probable by a study of hard X-ray loop images by Švestka and Poletto (1985) who concluded that the behavior of the loops could be explained by magnetic reconnection of previously distended field lines; see also Kopp and Poletto (1984). In this picture, the destabilization of the magnetic field also initiates the CME, which therefore does not causally follow the flare. Consequently, one may get the impression that the flare and the CME are independent phenomena, while the correct interpretation is that they are initiated separately but are due to the same overall magnetic realignment. Simnett and Harrison (1985) have taken this idea a step further and developed a scenario which starts with an energy release as advocated by Heyvaerts *et al.* (1977) (see Section 7.1.6). The shock waves produced by the energy release will accelerate electrons and protons, and these again will heat the lower coronal gas and trigger a magnetically driven CME. In this scenario further changes in the magnetic field, as the CME is driven out, result in a fresh energy release which is identified as the flare. It is known from observations (see Section 6.5.2) that up to several minutes before the impulsive phase of a flare, a so-called flare onset (precursor) can be discerned. One observes an increase in high-energy emission ($E < 200\,\text{keV}$) in the form of hard X-rays and radio waves, as well as a broadening of soft X-ray emission lines, indicating that both heating and some particle acceleration take place in the early rearrangement of the magnetic field structure before the cataclysmic energy release in the impulsive phase.

Epilogue

Exit, pursued by a flare
The Winter's Tale, with apologies to W. Shakespeare (1564–1616)

In this short final chapter, we briefly outline the current status in our understanding of solar flare physics, acquired over the past century or so, and also describe what advances we can expect to see in the near future.

We do know that solar flares are explosive events occurring in the solar corona, transition region, and chromosphere, and that they have their origin in the sudden release of energy stored in stressed magnetic fields (Chapter 3). The details of this energy-release process are, however, highly uncertain – a formidable problem lies in extracting the energy from the magnetic field because of the very low resistivity of the solar atmosphere (Chapters 3 and 6). Much knowledge has been gained by analyzing the radiation signatures of the impulsive (Chapter 7) and gradual (Chapter 8) phases and in attempting to reconcile these various spectral signatures in terms of a self-consistent energy-transport model. In this way we hope to constrain the nature of the energy-release process itself, such as whether it has predominantly thermal or nonthermal effects (Chapter 7). Such studies involve the complex interplay of radiation transport theory (Chapter 5), magnetohydrodynamics (Chapter 3), plasma physics and kinetic theory (Chapter 4), and atomic and nuclear physics (Chapters 2 and 5). It is largely due to the complex nature of these interactions that the solar flare problem is so fascinating and challenging (not to mention difficult).

Our observational knowledge of solar flares has increased dramatically in the last few decades, as a result of the many successful satellite observatories and new ground-based facilities. During the last solar maximum especially, spacecraft such as the Solar Maximum Mission, P78-1, and Hinotori gained many qualitative as well as quantitative advances in our understanding of solar flares. The first spatial imaging of hard X-ray sources was accomplished with simultaneous imaging data at other wavelengths, such as soft X-ray, EUV, optical, and radio. High time resolution optical, radio, and X-ray data revealed that particle acceleration occurs rapidly during the impulsive phase, with particles of up to billion electron volt energies being accelerated in the first few seconds. Soft X-ray and γ-ray line spectra have enabled accurate determination of chemical abundances in flaring regions, and ground-based magnetographs have revealed the magnetic stress that leads to a solar flare.

Future observations currently being planned promise equally great advances. Using Fourier transform image reconstruction techniques we now have the capability to directly image, with a resolution of order 1 arc sec, hard X-rays at several hundred kiloelectron volts well above the level where background thermal emission is significant. Such observations will allow us to probe the nature of particle acceleration and transport processes on a scale commensurate with the physical size scales appropriate to these processes. Cooled germanium γ-ray spectrometers, with spectral resolution of the order of 1 keV, will allow us to perform true γ-ray spectroscopy, in the sense of measuring line profiles. This will help solar physicists understand the role of heavy particles and protons in the flare process. Multifrequency radio observations, at high spatial resolution, will be used to define the microwave spectra in flares and also the characteristics of the exciting electron population. Instruments capable of measuring statistically significant hard X-ray polarizations at high (≈ 100 keV) energies have now been developed, and a combination of different spacecraft over various heliocentric longitudes could measure directly the directivity of the hard X-ray emission. The interested reader can refer to the *MAX '91* report by Dennis *et al.* (1986) for more details of these and other plans for the solar maximum in 1990–95.

On the theoretical front, the advent of high-speed computers has enabled solar physicists to tackle problems of previously unmanageable complexity. Among these are the radiative transport in lines and continua throughout an atmosphere consisting of several hundred atomic species formed over a 3 order-of-magnitude range in temperature, the magnetohydrodynamics of energized flare atmospheres, and the collective plasma physics of an accelerated population of particles interacting with a background atmosphere. Scientists from other disciplines are becoming increasingly involved in the solar flare problem, and the resulting symbiotic exchange of information and ideas promises great advances in our theoretical understanding of energy release and transport in solar flares.

In conclusion then, the study of solar flares has not only an exciting past and present, but an exciting and challenging future as well. We the authors, hope that some of our own enthusiasm for the subject is apparent in this book, and that it will inspire people to themselves pursue solar flare physics research, with its unique blend of observations, modeling, and theory. In closing, we remind ourselves that, as remarked in the introduction, our observational knowledge of solar flares has progressed well ahead of a corresponding theoretical understanding. We are, however, cautioned by the words of Sir Arthur Eddington: 'It is also a good rule not to put too much confidence in experimental results until they have been confirmed by theory.'

Que ceci soit la fin du livre – mais non la fin de la recherche.

Saint Bernard (1098–1153)

References

A man will turn over half a library to make one book

S. Johnson (1709–84)

Agrawal, P. C., Rao, A. R. and Riegler, G. R. (1986a). *Mon. Not. Roy. Astr. Soc.*, **219**, 777.

Agrawal, P. C., Rao, A. R. and Sreekantan, B. V. (1986b). *Mon. Not. Roy. Astr. Soc.*, **219**, 225.

Alam, B. and Ansari, S. M. R. (1985). *Solar Phys.*, **96**, 219.

Alfvén, H. (1950). *Cosmical Electrodynamics*, Clarendon Press, Oxford.

Alfvén, H. (1981). *Cosmic Plasma*, Reidel, Dordrecht.

Alfvén, H. (1987). In *Double Layers in Astrophysics* (Williams, A. C. and Moorehead, T. W., eds.), p. 1, NASA Conf. Publ. 2469.

Alfvén, H. and Carlqvist, P. (1967). *Solar Phys.*, **1**, 220.

Allen, C. W. (1973). *Astrophysical Quantities*, 3rd edn, Athlone Press, London.

Anderson, G. F. (1966). Ph.D. Thesis, University of Colorado, Boulder.

Antiochos, S. K. (1980). *Astrophys. J.*, **236**, 270.

Antiochos, S. K. and Sturrock, P. A. (1976). *Solar Phys.*, **49**, 359.

Antiochos, S. K. and Sturrock, P. A. (1978). *Astrophys. J.*, **220**, 1137.

Antonucci, E. and Dennis, B. R. (1983). *Solar Phys.*, **86**, 67.

Antonucci, E., Gabriel, A. H., Acton, L. W., Culhane, J. L., Doyle, J. G., Leibacher, J. W., Machado, M. E., Orwig, L. E. and Rapley, C. G. (1982). *Solar Phys.*, **78**, 107.

Antonucci, E., Gabriel, A. H. and Dennis, B. R. (1984). *Astrophys. J.*, **287**, 917.

Anzer, U. (1978). *Solar Phys.*, **57**, 111.

Athay, R. G. (1984). *Solar Phys.*, **93**, 123.

Athay, R. G. and Moreton, G. E. (1961). *Astrophys. J.*, **133**, 935.

Athay, R. G., White, O. R., Lites, B. W. and Bruner, E. C. (1980). *Solar Phys.*, **66**, 357.

Avignon, Y. and Pick-Gutmann, M. (1959). *Compt. Rend.*, **248**, 368.

Babcock, H. D. and Babcock, H. W. (1953). In *The Sun* (Kuiper, G., ed.), p. 704, University of Chicago Press, Chicago.

Babcock, H. D. and Babcock, H. W. (1955). *Astrophys. J.*, **121**, 349.

Bai, T. (1979). *Solar Phys.*, **62**, 113.

Bai, T. (1982). *Astrophys. J.*, **259**, 341.

Bai, T. and Ramaty, R. (1978). *Astrophys. J.*, **219**, 705.

Bai, T., Hudson, H. S., Pelling, R. M., Lin, R. P., Schwartz, R. A. and von Rosenvinge, T. T. (1983). *Astrophys. J.*, **267**, 433.

Basri, G. S., Linsky, J. L., Bartoe, J.-D. F., Brueckner, G. and van Hoosier, M. E. (1979). *Astrophys. J.*, **230**, 924.

Batchelor, D. A., Crannell, C. J., Wiehl, N. J. and Magun, A. (1985). *Astrophys. J.*, **295**, 258.

Becker, U. (1957). *Z. Ap.*, **42**, 85.

Beckers, J. M. and Schröter, E. H. (1968). *Solar Phys.*, **4**, 142.

Bekefi, G. (1966). *Radiative Processes in Plasmas*, John Wiley & Sons, Inc., New York.

Bell, B. and Glazer, H. (1959). *Smithsonian Contributions to Astrophysics*, **3**, 25.

Bely-Dubau, F. and Gabriel, A. H. (1984). *Mem. S. A. It.*, **55**(4), 685.

Bely-Dubau, F., Gabriel, A. H. and Volonté, S. (1979a). *Mon. Not. Roy. Astr. Soc.*, **186**, 405.

Bely-Dubau, F., Gabriel, A. H. and Volonté, S. (1979b). *Mon. Not. Roy. Astr. Soc.*, **189**, 802.

Bely-Dubau, F., Dubau, J., Faucher, P. and Gabriel, A. H. (1982a). *Mon. Not. Roy. Astr. Soc.*, **198**, 239.

Bely-Dubau, F., Dubau, J., Faucher, P., Gabriel, A. H., Loulerque, M., Steenman-Clark, L., Volonté, S., Antonucci, E. and Rapley, C. G. (1982b). *Mon. Not. Roy. Astr. Soc.*, **201**, 1155.

Benz, A. O. (1985). *Solar Phys.*, **96**, 357.

Benz, A. O. (1987). In *Rapid Fluctuations in Solar Flares* (Dennis, B. R., Orwig, L. E. and Kiplinger, A. L., eds.), p. 133, NASA Conf. Publ. 2449.

Benz, A. O. and Wentzel, D. G. (1981). *Astron. Astrophys.*, **94**, 100.

Bhalla, C. P., Gabriel, A. H. and Presmyakov, L. P. (1975). *Mon. Not. Roy. Astr. Soc.*, **172**, 359.

Bhatnagar, A. (1986). In *Solar Terrestrial Physics, Indo-US 1984 Workshop Proceedings* (Kundu, M. R., Biswas, B., Reddy, B. M. and Ramadurai, S., eds.), p. 73, National Physical Laboratory, New Delhi.

Billings, D. E. (1966). *A Guide to the Solar Corona*, Academic Press, New York.

Boischot, A. (1957). *Compt. Rend.*, **244**, 1326.

Boischot, A. (1958). *Ann. Astrophys.*, **21**, 273.

Born, M. and Wolf, E. (1965). *Principles of Optics*, Pergamon Press, New York.

Borovsky, J. E. (1983). *Phys. Fluids*, **36**, 3273.

Bray, R. J. and Loughhead, R. E. (1965). *Sunspots*, John Wiley & Sons, Inc., New York.

Breit, G. (1925). *J. Opt. Soc. Am.*, **10**, 439.

Breit, G. (1933). *Rev. Mod. Phys.*, **5**, 91.

Brown, D. G. and Emslie, A. G. (1987). *Solar Phys.*, **110**, 305.

Brown, J. C. (1971). *Solar Phys.*, **18**, 489.

Brown, J. C. (1972). *Solar Phys.*, **26**, 441.

Brown, J. C. (1973). *Solar Phys.*, **31**, 143.

Brown, J. C. (1974). In *Coronal Disturbances*, Proc. IAU Symp. 57 (Newkirk, Jr., G. A., ed.), p. 105, Reidel, Dordrecht.

Brown, J. C. (1975). In *Solar Gamma-, X-, and EUV Radiation*, Proc. IAU Symp. 68 (Kane, S. R., ed.), p. 305, Reidel, Dordrecht.

Brown, J. C. (1978). *Astrophys. J.*, **225**, 1076.

Brown, J. C. and Bingham, R. (1984). *Astron. Astrophys.*, **131**, L11.

Brown, J. C. and Emslie, A. G. (1988). *Astrophys. J.*, **333**, in press.

Brown, J. C. and MacKinnon, A. L. (1985). *Astrophys. J. Lett.*, **292**, L31.

Brown, J. C., Hayward, J. and Spicer, D. S. (1981). *Astrophys. J. Lett.*, **245**, L91.

Brown, J. C., Melrose, D. B. and Spicer, D. S. (1979). *Astrophys. J.*, **228**, 592.

Brueckner, G. E. (1980). *Highlights of Astronomy*, **5**, 557.

Bruner, Jr., E. C. and Lites, B. W. (1979). *Astrophys. J.*, **228**, 322.

Bruzek, A. (1951). *Z. Ap.*, **13**, 277.

Bruzek, A. (1957). *Z. Ap.*, **42**, 76.

Bumba, V. (1958). *Izv. Crim. Astrophys. Obs.*, **19**, 105.

Bumba, V. (1986). *Bull. Astron. Inst. Czechosl.*, **37**, 281.

Burgess, A. (1964). *Astrophys. J.*, **139**, 776.

Burgess, A. and Seaton, M. (1960). *Mon. Not. Roy. Astr. Soc.*, **120**, 121.

Burlaga, L. F. (1971). *Space Sci. Rev.*, **12**, 600.

Burlaga, L. F. and Ogilvie, K. W. (1969). *J. Geophys. Res.*, **74**, 2815.

Burlaga, L. F., Sittler, E., Mariani, F. and Schwenn, R. (1981). *J. Geophys. Res.*, **86**, 6673.

Campbell, P. M. (1984). *Phys. Rev. A*, **30**, 365.

Canfield, R. C. *et al.* (1979). In *Solar Flares – A Monograph from Skylab Solar Workshop II* (Sturrock, P. A., ed.), Chapter 6, Colorado Associated University Press, Boulder.

Canfield, R. C. and Gayley, K. G. (1987). *Astrophys. J.*, **322**, 999.

Canfield, R. C., Gunkler, T. A. and Ricchiazzi, P. J. (1984). *Astrophys. J.*, **282**, 296.

Carmichael, H. (1964). In *AAS-NASA Symposium on Physics of Solar Flares* (Hess, W. N., ed.), p. 451, NASA SP-50.

Carrington, R. C. (1859). *Mon. Not. Roy. Astr. Soc.*, **20**, 13.

Chambe, G. and Hénoux, J.-C. (1979). *Astron. Astrophys.*, **80**, 123.

Chanan, G. A., Emslie, A. G. and Novick, R. (1988). *Solar Phys.*, in press.

Chandrasekhar, S. (1950). *Radiative Transfer*, Clarendon Press, Oxford.

Chandrasekhar, S. and Kendall, P. C. (1957). *Astrophys. J.*, **126**, 457.

Chandrashekar, S. and Emslie, A. G. (1987). *Solar Phys.*, **107**, 83.

Chapman, S. and Cowling, T. G. (1952). *Mathematical Theory of Non-Uniform Gases*, 2nd edn, Cambridge University Press.

Chen, F. F. (1974). *Principles of Plasma Physics*, McGraw-Hill, New York.

Cheng, C.-C., Pallavicini, R., Acton, L. W. and Tandberg-Hanssen, E. (1985). *Astrophys. J.*, **298**, 887.

Cheng, C.-C., Tandberg-Hanssen, E. and Orwig, L. E. (1984). *Astrophys. J.*, **278**, 853.

Chiuderi, C., Einandi, G., Ma, S. S. and van Hoven, G. (1980). *J. Plasma Phys.*, **24**, 39.

Chiuderi-Drago, F., Mein, N. and Pick, M. (1986). *Solar Phys.*, **103**, 235.

Chubb, T. A., Friedman, H. and Kreplin, R. W. (1966). *J. Geophys. Res.*, **71**, 3611.

Chupp, E. L., Forrest, D. J., Ryan, J. M., Heslin, J., Reppin, C., Pinkau, K., Kanbach, G., Rieger, E. and Share, G. N. (1982). *Astrophys. J. Lett.*, **263**, L95.

Clerke, A. M. (1903). *A Popular History of Astronomy*, A&C Black, London.

Condon, E. V. and Shortley, G. H. (1953). *The Theory of Atomic Spectra*, Cambridge University Press.

Craig, I. J. D. and Brown, J. C. (1976). *Astron. Astrophys.*, **49**, 239.

Craig, I. J. D. and Brown, J. C. (1985). *Inversion Problems in Astrophysics*, Adam Hilger, London.

Craig, I. J. D. and McClymont, A. N. (1976). *Solar Phys.*, **50**, 133.

Crannell, C. J., Frost, K. J., Matzler, C., Ohki, K. and Sada, J. L. (1978). *Astrophys. J.*, **223**, 620.

Crooker, N. V. (1983). In *Solar Wind Five* (Neugebauer, M., ed.), p. 303, NASA Conf. Publ. 2280.

Culhane, J. L. and Acton, L. W. (1970). *Mon. Not. Roy. Astr. Soc.*, **151**, 141.

Culhane, J. L. *et al.* (1981). *Astrophys. J. Lett.*, **244**, L141.

Datlowe, D. W., O'Dell, S. L., Peterson, L. E. and Elcan, M. J. (1977). *Astrophys. J.*, **212**, 561.

de Groot, T. (1962). *Inf. Bull. Solar Radio Obs.* (Europe), **9**, 3.

de Jager, C. and de Jonge, G. (1978). *Solar Phys.*, **58**, 127.

de Jager, C. and Švestka, Z. (1985). *Solar Phys.*, **100**, 435.

Dennis, B. R. (1982). *OSO-8 Internal Technical Memorandum*, Goddard Space Flight Center, Maryland.

Dennis, B. R. (1985). *Solar Phys.*, **100**, 465.

Dennis, B. R. *et al.* (1986). *MAX '91, An Advanced Payload for the Exploration of High Energy Processes on the Active Sun*, Report of MAX '91 Science Study Committee, NASA, Washington, D.C.

Dere, K. P., Horan, D. M. and Kreplin, R. W. (1974). *J. Atmos. Terr. Phys.*, **36**, 989.

Dodson-Prince, H. W. and Hedeman, E. R. (1960). *Astron. J.*, **65**, 51.

Dodson-Prince, H. W. and Hedeman, E. R. (1964). In *AAS-NASA Symposium on Physics of Solar Flares* (Hess, W. N., ed.), p. 15, NASA SP-50.

Dodson-Prince, H. W. and Hedeman, E. R. (1970). *Solar Phys.*, **13**, 401.

Donati-Falchi, A., Falciani, R. and Smaldone, L. A. (1985). *Astron. Astrophys.*, **152**, 165.

Donnelly, R. F. (1976). *J. Geophys. Res.*, **81**, 4745.

Donnelly, R. F. and Kane, S. R. (1978). *Astrophys. J.*, **222**, 1043.

Doschek, G. A. (1985). In *Autoionization* (Temkin, A., ed.), Chapter 6, Plenum Publishing Co., New York.

Doschek, G. A., Feldman, U. and Cowan, R. D. (1981). *Astrophys. J.*, **245**, 315.

Doschek, G. A., Feldman, U. and Kreplin, R. W. (1980). *Astrophys. J.*, **239**, 725.

Dröge, F. and Riemann, P. (1961). *Inf. Bull. Solar Radio Obs.* (Europe), **8**, 6.

Dryer, M. (1974). *Space Sci. Rev.*, **15**, 403.

Dryer, M. (1982). *Space Sci. Rev.*, **33**, 233.

Dryer, M. and Jones, D. L. (1968). *J. Geophys. Res.*, **73**, 4875.

Dryer, M. and Maxwell, A. (1979). *Astrophys. J.*, **231**, 945.

Dryer, M., Pérez-de-Tejada, H., Taylor, Jr., H. A., Intriligator, D. S., Mihalov, J. D. and Rompolt, B. (1982). *J. Geophys. Res.*, **87**, 9035.

Dryer, M., Wu, S. T., Steinolfson, R. S. and Wilson, R. M. (1979). *Astrophys. J.*, **227**, 1059.

Dubau, J., Gabriel, A. H., Loulerque, M., Steenman-Clark, L. and Volonté, S. (1981). *Mon. Not. Roy. Astr. Soc.*, **195**, 705.

Duijveman, A., Hoyng, P. and Machado, M. E. (1982). *Solar Phys.*, **81**, 137.

Dulk, G. A. (1985). *Ann. Rev. Astron. Astrophys.*, **23**, 169.

Dulk, G. A. and Marsh, K. A. (1982). *Astrophys. J.*, **259**, 350.

Dulk, G. A., Melrose, D. B. and White, S. M. (1979). *Astrophys. J.*, **234**, 1137.

Duncan, R. A. (1981). *Solar Phys.*, **73**, 191.

D'Uston, C., Dryer, M., Han, S. M. and Wu, S. T. (1981). *J. Geophys. Res.*, **86**, 525.

Elgaröy, Ö. (1961). *Astrophys. Norw.*, **7**, 4.

Elgaröy, Ö. (1962). *Inf. Bull. Solar Radio Obs.* (Europe), **9**, 4.

Elgaröy, Ö. (1977). *Solar Noise Storms*, Pergamon Press, Oxford.

Ellison, M. A., McKenna, S. M. P. and Reid, J. H. (1961). *Mon. Non.*, **122**, 491.

Emslie, A. G. (1978). *Astrophys. J.*, **224**, 241.

Emslie, A. G. (1980). *Astrophys. J.*, **235**, 1055.

Emslie, A. G. (1981). *Astrophys. J.*, **245**, 711.

Emslie, A. G. (1983). *Solar Phys.*, **86**, 133.

Emslie, A. G. (1985). *Solar Phys.*, **98**, 281.

Emslie, A. G. (1986). In *Solar Flares and Coronal Physics Using P/OF as a Research Tool* (Tandberg-Hanssen, E., Wilson, R. M. and Hudson, H. S., eds.), p. 132, NASA Conf. Publ. 2421.

Emslie, A. G. and Alexander, D. (1987). *Solar Phys.*, **110**, 295.

Emslie, A. G. and Brown, J. C. (1980). *Astrophys. J.*, **237**, 1015.

Emslie, A. G. and Machado, M. E. (1979). *Solar Phys.*, **64**, 129.

Emslie, A. G. and Machado, M. E. (1987). *Solar Phys.*, **107**, 263.

Emslie, A. G. and Nagai, F. (1985). *Astrophys. J.*, **288**, 779.

Emslie, A. G. and Noyes, R. W. (1978). *Solar Phys.*, **57**, 373.

Emslie, A. G. and Rust, D. M. (1979). *Solar Phys.*, **65**, 271.

Emslie, A. G. and Smith, D. F. (1984). *Astrophys. J.*, **279**, 382.

Emslie, A. G. and Sturrock, P. A. (1981). *Solar Phys.*, **80**, 99.

Emslie, A. G. and Vlahos, L. (1980). *Astrophys. J.*, **242**, 359.

Emslie, A. G., Brown, J. C. and Donnelly, R. F. (1978). *Solar Phys.*, **57**, 175.

Emslie, A. G., Brown, J. C. and Machado, M. E. (1981). *Astrophys. J.*, **246**, 337.

Emslie, A. G., Fennelly, J. A. and Machado, M. E. (1986a). *Adv. Space Res.*, **6**(6), 139.

Emslie, A. G., Phillips, K. J. H. and Dennis, B. R. (1986b). *Solar Phys.*, **103**, 89.

Falciani, R., Giordano, M., Rigutti, M. and Roberti, G. (1977). *Solar Phys.*, **54**, 169.

Fárník, F., Kaastra, J., Kálmán, B., Karlický, M., Slottje, C. and Valnicek, B. (1983). *Solar Phys.*, **89**, 355.

Feldman, U. (1980). *Physica Scripta*, **24**, 681.

Feldman, U., Doschek, G. A., Kreplin, R. W. and Mariska, J. T. (1980). *Astrophys. J.*, **241**, 1175.

Ferraro, V. C. A. and Plumpton, C. (1966). *An Introduction to Magneto-Fluid Mechanisms*, Clarendon Press, Oxford.

Fisher, G. H. (1987). *Astrophys. J.*, **317**, 502.

Fisher, G. H., Canfield, R. C. and McClymont, A. N. (1985*a*). *Astrophys. J.*, **289**, 414.

Fisher, G. H., Canfield, R. C. and McClymont, A. N. (1985*b*). *Astrophys. J.*, **289**, 425.

Fisk, L. A. (1978). *Astrophys. J.*, **224**, 1048.

Forrest, D. J. and Chupp, E. L. (1983). *Nature*, **305**, 291.

Fried, B. D. and Gould, R. W. (1961). *Phys. Fluids*, **4**, 139.

Furth, H. P., Killeen, J. and Rosenbluth, M. N. (1963). *Phys. Fluids*, **6**, 459.

Gabriel, A. H. (1972). *Mon. Not. Roy. Astr. Soc.*, **160**, 99.

Gabriel, A. H. and Jordan, C. (1972). In *Case Studies in Atom. Coll. Physics II* (McDaniel, E. W. and McDowell, M. R. C., eds.), Chapter 4, pp. 209–91, North-Holland Publishing Co., Amsterdam.

Gabriel, A. H. *et al.* (1981). *Astrophys. J. Lett.*, **244**, L147.

Gaizauskas, V., Harvey, K. L., Harvey, J. W. and Zwaan, C. (1983). *Astrophys. J.*, **265**, 1056.

Gary, D. E. (1985). *Astrophys. J.*, **297**, 799.

Gary, D. E., Dulk, G. A., House, L. L., Illing, R. M. E., Sawyer, C., Wagner, W. J., McLean, D. J. and Hildner, E. (1984). *Astron. Astrophys.*, **134**, 222.

Gary, D. E., Dulk, G. A., House, L. L., Illing, R., Wagner, W. J. and McLean, D. J. (1985). *Astron. Astrophys.*, **152**, 42.

Gaunt, J. A. (1930). *Phil. Trans.*, **229**, 163.

Geltman, S. (1985). *J. Phys. B: At. Mol. Phys.*, **18**, 1425.

Gergely, T. E. (1984). In *STIP Symposium on Solar/Interplanetary Intervals* (Shea, M. A., Smart, D. F. and McKenna-Lawlor, S. M. P., eds.), p. 347, Book Crafters, Inc., Chelsea, Michigan.

Gergely, T. E. (1986). In *Solar Terrestrial Physics, Indo-US Workshop 1984 Proceedings* (Kundu, M. R., Biswas, B., Reddy, B. M. and Ramadurai, S., eds.), p. 303, National Physical Laboratory, New Delhi.

Gergely, T. E., Kundu, M. R., Erskine, III, F. T., Sawyer, C., Wagner, W. J., Illing, R. M. E., House, L. L., McCabe, M. K., Stewart, R. T., Nelson, G. J., Koomen, M. J., Michels, D., Howard, R. and Sheeley, N. (1984). *Solar Phys.*, **90**, 161.

Ginzburg, V. L. and Syrovatskii, S. I. (1965). *Ann. Rev. Astron. Astrophys.*, 3, 297.

Giovanelli, R. G. (1939). *Astrophys. J.*, **89**, 555.

Glackin, D. L. (1974). *Solar Phys.*, **36**, 51.

Gluckstern, R. L. and Hull, M. N. (1953). *Phys. Rev.*, **90**, 1030.

Gold, T. (1964). In *AAS-NASA Symposium on Physics of Solar Flares* (Hess, W. N., ed.), p. 389, NASA SP-50.

Gold, T. and Hoyle, F. (1960). *Mon. Not. Roy. Astr. Soc.*, **120**, 89.

Golub, L., Krieger, A. S., Harvey, J. W. and Vaiana, G. S. (1977). *Solar Phys.*, **53**, 111.

Golub, L., Krieger, A. S., Silk, J. K., Timothy, A. F. and Vaiana, G. S. (1974). *Astrophys. J. Lett.*, **189**, L93.

Gosling, J. T., Asbridge, J. R., Bame, S. L., Hundhausen, A. J. and Strong, I. B. (1967). *J. Geophys. Res.*, **72**, 3357.

Gosling, J. T., Hildner, E., MacQueen, R. M., Munro, R. H., Poland, A. I. and Ross, C. L. (1976). *Solar Phys.*, **48**, 389.

Griem, H. (1964). *Plasma Spectroscopy*, McGraw-Hill, New York.

Grossi-Gallegos, H., Molnar, M. and Seibold, J. R. (1971). *Solar Phys.*, **16**, 120.

Hagyard, M. J., Smith, Jr., J. B., Teuber, D. and West, E. A. (1984). *Solar Phys.*, **91**, 115.

Haisch, B. M. and Linsky, J. L. (1980). *Astrophys. J. Lett.*, **236**, L33.

Hanle, W. (1923). *Naturwiss.*, **11**, 691.

Hanle, W. (1924). *Z. Phys.*, **29–30**, 93.

Hanle, W. (1925). *Naturwiss.*, **14**, 214.

Haurwitz, M. W., Yoshida, S. and Akasofu, S.-I. (1965). *J. Geophys. Res.*, **70**, 2977.

Heisenberg, W. (1926). *Z. Phys.*, **31**, 617.

Heitler, W. (1954). *The Quantum Theory of Radiation*, Oxford University Press.

Hénoux, J.-C. (1975). *Solar Phys.*, **42**, 219.

Hénoux, J.-C. (1983). In *Proceedings of the Japan-France Seminar on Active Phenomena in the Outer Atmosphere of the Sun and Stars* (Pecker, J.-C. and Uchida, Y., eds.), p. 200, Observatoire de Paris-Meudon, France.

Hénoux, J.-C. and Rust, D. (1980). *Astron. Astrophys.*, **91**, 322.

Hernandez, A. M., Machado, M. E., Vilmer, N. and Trottet, G. (1986). *Astron. Astrophys.*, **167**, 77.

Heyvaerts, J. (1974). *Solar Phys.*, **38**, 419.

Heyvaerts, J., Priest, E. R. and Rust, D. M. (1977). *Astrophys. J.*, **216**, 123.

Hiei, E. (1982). *Solar Phys.*, **80**, 113.

Hildner, E. (1977). In *Study of Travelling Interplanetary Phenomena* (Shea, M. A., Smart, D. F. and Wu, S. T., eds.), p. 3, Reidel, Dordrecht.

Hildner, E., Gosling, J. T., MacQueen, R. M., Munro, R. H., Poland, A. I. and Ross, C. L. (1975). *Solar Phys.*, **42**, 163.

Holman, G. D. (1985). *Astrophys. J.*, **293**, 584.

Holtsmark, J. (1919). *Ann. Physik*, **58**, 577.

House, L. L. (1971). *J. Quant. Spectr. Rad. Transf.*, **11**, 367.

House, L. L. (1970a). *J. Quant. Spectr. Rad. Transf.*, **10**, 909.

House, L. L. (1970b). *J. Quant. Spectr. Rad. Transf.*, **10**, 1171.

Howard, R. A., Michels, D. J., Sheeley, Jr., N. R. and Koomen, M. J. (1982). *Astrophys. J. Lett.*, **263**, L101.

Howard, R. A., Sheeley, Jr., N. R., Koomen, M. J. and Michels, D. J. (1985). *J. Geophys. Res.*, **90**, 8173.

Hoyng, P., Brown, J. C. and van Beek, H. F. (1976). *Solar Phys.*, **48**, 197.

Hoyng, P., Duijveman, A., Machado, M. E., Rust, D. M., Švestka, Z., Boelee, A., de Jager, C., Frost, K. J., Lafleur, H., Simnett, G. M., van Beek, H. F. and Woodgate, B. E. (1981). *Astrophys. J. Lett.*, **246**, L155.

Hudson, H. S. (1972). *Solar Phys.*, **24**, 414.

Hudson, H. S. (1973). In *Symposium on High Energy Phenomena on the Sun* (Ramaty, R. and Stone, R. G., eds.), p. 209, NASA SP-342.

Hundhausen, A. J. (1972a). In *Solar-Terrestrial Physics Part II: The Interplanetary Medium* (Dryer, E. R. and Roederer, J. G., eds.), p. 1, Reidel, Dordrecht.

Hundhausen, A. J. (1972b). *Coronal Expansion and Solar Wind*, Springer-Verlag, New York.

Hundhausen, A. J. (1987). *J. Geophys. Res.*, submitted.

Hundhausen, A. J., Holzer, T. E. and Low, B. C. (1987). *J. Geophys. Res.*, **92**, 11,173.

Hundhausen, A. J., MacQueen, R. M. and Sime, D. G. (1984). *Eos Trans. AGU*, **65**, 1069.

Hyder, C. L. (1967a). *Solar Phys.*, **2**, 49.

Hyder, C. L. (1967b). *Solar Phys.*, **2**, 267.

Hyder, C. L. (1968). In *Mass Motions in Solar Flares and Related Phenomena*, Nobel Symp. 9 (Ohman, Y., ed.), p. 57, Almqvist & Wiksell, Stockholm.

IAU (1966). 'Report of the Working Committee of Commission 10, Commission de l'Activité Solaire', *Trans. IAU XII General Assembly*, p. 151, Reidel, Dordrecht.

Inglis, D. R. and Teller, E. (1939). *Astrophys. J.*, **90**, 439.

Ionson, J. A. (1982). *Astrophys. J.*, **254**, 318.

Jackson, B. V. (1979). *Proc. Astr. Soc. of Australia*, **3**, 383.

Jackson, B. V. and Hildner, E. (1978). *Solar Phys.*, **60**, 155.

Jackson, B. V., Dulk, G. A. and Sheridan, K. V. (1980). In *Solar and Interplanetary Dynamics*, Proc. IAU Symp. 91 (Dryer, M. and Tandberg-Hanssen, E., eds.), p. 279, Reidel, Dordrecht.

Jackson, J. D. (1962). *Classical Electrodynamics*, John Wiley & Son, Inc., New York.

Jefferies, J. T. and Thomas, R. N. (1958). *Astrophys. J.*, **127**, 667.

Jokipii, J. R. (1966). *Astrophys. J.*, **143**, 961.

Kahler, S. W. (1971a). *Astrophys. J.*, **164**, 365.

Kahler, S. W. (1971b). *Astrophys. J.*, **168**, 319.

Kahler, S. W. (1978). *Solar Phys.*, **59**, 87.

Kahler, S. W., Lin, R. P., Reames, D. V., Stone, R. G. and Liggett, M. (1987). *Solar Phys.*, **107**, 385.

Kane, S. R. (1981). *Astrophys. J.*, **247**, 113.

Kane, S. R. and Donnelly, R. F. (1971). *Astrophys. J.*, **164**, 151.

Kane, S. R., Anderson, K. A., Evans, W. D., Klebesadel, R. W. and Laros, J. G. (1979a). *Astrophys. J. Lett.*, **233**, L151.

Kane, S. R., Anderson, K. A., Evans, W. D., Klebesadel, R. W. and Laros, J. G. (1980). *Astrophys. J. Lett.*, **239**, L85.

Kane, S. R., Frost, K. J. and Donnelly, R. F. (1979b). *Astrophys. J.*, **234**, 669.

Kane, S. R., Kai, K., Kosugi, T., Enome, S., Landecker, P. and McKenzie, D. L. (1986). *Astrophys. J.*, **271**, 376.

Kane, S. R., Love, J. J., Neidig, D. F. and Cliver, E. W. (1985). *Astrophys. J. Lett.*, **290**, L45.

Kaufmann, P., Correia, E., Costa, J. E. R., Dennis, B. R., Hurford, G. J. and Brown, J. C. (1984). *Solar Phys.*, **91**, 359.

Kaufmann, P., Correia, E., Costa, J. E. R., Sawant, H. S. and Zodi, A. M. (1985). *Solar Phys.*, **95**, 155.

Keenan, F. P., Kingston, A. E. and McKenzie, D. L. (1985). *Astrophys. J.*, **291**, 855.

Kiplinger, A. L. (1986). Personal communication.

Kiplinger, A. L., Dennis, B. R., Emslie, A. G., Frost, K. J. and Orwig, L. E. (1983). *Astrophys. J. Lett.*, **265**, L99.

Kleczek, J. (1963). In *The Solar Corona* (Evans, J., ed.), p. 151, Academic Press, New York.

Klein, K.-L., Pick, M., Magun, A. and Dennis, B. R. (1987). *Solar Phys.*, **111**, 225.

Klein, L. W. and Burlaga, L. F. (1982). *J. Geophys. Res.*, **87**, 613.

Kniffen, D. A., Fichtel, C. E. and Thompson, D. J. (1977). *Astrophys. J.*, **215**, 765.

Knight, J. W. and Sturrock, P. A. (1977). *Astrophys. J.*, **218**, 306.

Koch, H. W. and Motz, J. W. (1959). *Rev. Mod. Phys.*, **31**, 920.

Kopp, R. A. and Pneuman, G. (1976). *Solar Phys.*, **50**, 85.

Kopp, R. A. and Poletto, G. (1984). In *Proc. IAU Colloquium 86 on UV and X-ray Spectroscopy of Astrophysical and Laboratory Plasmas* (Doschek, G. A., ed.), p. 17, Reidel, Dordrecht.

Kovács, Á. and Dezsö, L. (1986). *Adv. Space Res.*, **6**(6), 29.

Krall, N. A. and Trivelpiece, A. W. (1973). *Principles of Plasma Physics*, McGraw-Hill, New York.

Krall, K. R., Smith, Jr., J. B., Hagyard, M. J., West, E. A. and Cumings, N. P. (1982). *Solar Phys.*, **79**, 59.

Kramers, H. A. (1923). *Phil. Mag.*, **46**(6), 836.

Kruskal, M. and Bernstein, I. B. (1964). *Phys. Fluids*, **7**, 407.

Kundu, M. R. (1965). *Solar Radio Astronomy*, Interscience Press, New York.

Kundu, M. R. (1982). *Rep. Prog. Phys.*, **45**, 1435.

Kundu, M. R. (1983a). *Solar Phys.*, **86**, 205.

Kundu, M. R. (1983b). *Adv. Space Res.*, **2**(11), 159.

Kundu, M. R. and Alissandrakis, C. E. (1984). *Solar Phys.*, **94**, 249.

Kundu, M. R. and Lang, K. R. (1985). *Science*, **228**, 9.

Kundu, M. R. and Vlahos, L. (1982). *Space Sci. Rev.*, **32**, 405.

Kundu, M. R. and Woodgate, B. (eds.) (1986). *Energetic Phenomena on the Sun, The Solar Maximum Mission Flare Workshop Proceedings*, NASA Conf. Publ. 2439.

Kundu, M. R., Erickson, W. E., Jackson, P. D. and Fainberg, J. (1970). *Solar Phys.*, **14**, 394.

Kundu, M. R., Gaizauskas, V., Woodgate, B. E., Schmahl, E. J., Shine, R. A. and Jones, H. P. (1985). *Astrophys. J. Suppl.*, **57**, 621.

Kundu, M. R., Gergely, T. E. and Golub, L. (1980). *Astrophys. J. Lett.*, **236**, L87.

Kundu, M. R., Machado, M. E., Erskine, F. T., Rovira, M. G. and Schmahl, E. J. (1984). *Astron. Astrophys.*, **132**, 241.

Kunkel, W. E. (1973). *Astrophys. J. Suppl.*, **25**, 1.

Kurochka, L. N. and Maslennikova, L. B. (1970). *Solar Phys.*, **11**, 33.

Landi Degl'Innocenti, E. (1982). *Solar Phys.*, **79**, 291.

Lang, K. R. (1986). In *Solar Terrestrial Physics, Indo-US Workshop 1984 Proceedings* (Kundu, M. R., Biswas, B., Reddy, B. M. and Ramadurai, S., eds.), p. 21, National Physical Laboratory, New Delhi.

Lang, K. R., Wilson, R. F. and Gaizauskas, V. (1983). *Astrophys. J.*, **267**, 455.

Lantos, P. (1985). In *Proc. of Kunming Workshop on Solar Physics and Interplanetary Travelling Phenomena* (de Jager, C. and Biao, C., eds.), p. 1082, Science Press, Beijing, China.

Lantos, P., Kerdraon, A., Rapley, C. G. and Bentley, R. D. (1981). *Astron. Astrophys.*, **101**, 33.

LaRosa, T. N. (1988). *Astrophys. J.*, in press.

Leach, J. and Petrosian, V. (1981). *Astrophys. J.*, **251**, 781.

Leach, J. and Petrosian, V. (1983). *Astrophys. J.*, **269**, 715.

Leach, J., Emslie, A. G. and Petrosian, V. (1985). *Solar Phys.*, **96**, 331.

Leblanc, T., Poquerousse, M. and Aubier, M. G. (1983). *Astron. Astrophys.*, **123**, 307.

Lemaire, P., Chouq-Bruston, M. and Vial, J.-C. (1984). *Solar Phys.*, **90**, 63.

Leroy, J.-L., Ratier, G. and Bommier, V. (1977). *Astron. Astrophys.*, **54**, 811.

Leroy, M. M. and Mangeney, A. (1984). *Annales Geophysicae*, **2**, 449.

Lin, R. P. (1974). *Space Sci. Rev.*, **16**, 184.

Lin, R. P. and Hudson, H. S. (1976). *Solar Phys.*, **50**, 153.

Lin, R. P., Schwartz, R. A., Kane, S. R., Pelling, R. M. and Hurley, K. C. (1984). *Astrophys. J.*, **283**, 421.

Lindholm, E. (1942). Ph.D. Thesis, Uppsala University, Sweden.

Lites, B. W. and Hansen, E. R. (1977). *Solar Phys.*, **55**, 347.

Lorentz, H. A. (1905a). *Proc. Roy. Acad.*, **14**, 518.

Lorentz, H. A. (1905b). *Proc. Roy. Acad.*, **14**, 577.

Low, B. C. (1982). *Rev. Geophys. Space Phys.*, **20**, 145.

Lyons, L. and Williams, D. (1985) (eds.). *Quantitative Aspects of Magnetospheric Physics*, Reidel, Dordrecht.

Machado, M. E. (1978). *Solar Phys.*, **60**, 341.

Machado, M. E. and Emslie, A. G. (1979). *Astrophys. J.*, **232**, 903.

Machado, M. E. and Linsky, J. L. (1975). *Solar Phys.*, **42**, 395.

Machado, M. E. and Rust, D. M. (1974). *Solar Phys.*, **38**, 499.

Machado, M. E., Avrett, E. H. and Emslie, A. G. (1988c). *Astrophys. J. Lett.*, in preparation.

Machado, M. E., Avrett, E. H., Vernazza, J. E. and Noyes, R. W. (1980). *Astrophys. J.*, **242**, 336.

Machado, M. E., Emslie, A. G. and Brown, J. C. (1978). *Solar Phys.*, **58**, 363.

Machado, M. E., Emslie, A. G. and Mauas, P. J. (1986). *Astron. Astrophys.*, **159**, 33.

Machado, M. E., Moore, R. L., Hernandez, A. M., Rovira, M. G., Hagyard, M. J. and Smith, Jr. J. B. (1988a). *Astrophys. J.*, **326**, 425–50.

Machado, M. E., Rovira, M. G. and Sneibrun, C. V. (1985). *Solar Phys.*, **99**, 189.

Machado, M. E., Xiao, Y. C., Wu, S. T., Prokakis, Th. and Dialetis, D. (1988b). *Astrophys. J.*, **326**, 451–61.

MacKinnon, A. L. (1985). *Solar Phys.*, **98**, 293.

MacKinnon, A. L., Brown, J. C. and Hayward, J. (1986). *Solar Phys.*, **99**, 231.

MacNeice, P., McWhirter, R. W. P., Spicer, D. S. and Burgess, A. (1984). *Solar Phys.*, **90**, 357.

MacNeice, P., Pallavicini, R., Mason, H. E., Simnett, G. M., Antonucci, E., Shine, R. A., Rust, D. M., Jordan, C. and Dennis, B. R. (1985). *Solar Phys.*, **99**, 167.

Makita, M., Hamana, S., Nishi, J., Shimizu, M., Koyamo, H., Sakurai, T. and Komatsu, H. (1985). *Publ. Astron. Soc. Japan*, **37**, 561.

Manheimer, W. M. (1977). *Phys. Fluids*, **20**, 265.

Manheimer, W. M. and Klein, H. H. (1975). *Phys. Fluids*, **18**, 1299.

Mariska, J. T. and Poland, A. I. (1985). *Solar Phys.*, **96**, 317.

Marsh, K. A. and Hurford, G. J. (1980). *Astrophys. J. Lett.*, **240**, L111.

Marsh, K. A. and Hurford, G. J. (1982). *Ann. Rev. Astron. Astrophys.*, **20**, 497.

Marsh, K. A., Hurford, G. J., Zirin, H., Dulk, G. A., Dennis, B. R., Frost, K. J. and Orwig, L. E. (1981). *Astrophys. J.*, **251**, 797.

Martin, S. F. and Ramsey, H. E. (1972). In *Solar Activity Observations and Predictions* (McIntosh, P. S. and Dryer, M., eds.), p. 371, MIT Press, Cambridge, Massachusetts.

Martin, S. F., Bentley, R. D., Schadee, A., Antalova, A., Kucera, A., Dezsö, L., Gesztelyi, L., Harvey, K. L., Jones, H., Livi, S. H. B. and Wang, J. (1984). *Adv. Space Res.*, **4**(7), 61.

Martin, S. F., Dezsö, L., Antalova, A., Kucera, A. and Harvey, K. L. (1982). *Adv. Space Res.*, **2**(11), 39.

Martres, M.-J. (1956). *L'Astronomie*, **70**, 401.

Martres, M.-J. and Pick, M. (1962). *Ann. Astrophys.*, **25**, 293.

Martres, M.-J., Michard, R. and Soru-Iscovici, I. (1966). *Ann. Astrophys.*, **29**, 249.

Martres, M.-J., Michard, R., Soru-Iscovici, I. and Tsap, T. T. (1968). *Solar Phys.*, **5**, 187.

Martres, M.-J., Mouradian, Z. and Soru-Escaut, I. (1986). *Astron. Astrophys.*, **161**, 376.

Martres, M.-J., Soru-Escaut, I. and Rayrole, J. (1971). In *Solar Magnetic Fields*, Proc. IAU Symp. 43 (Howard, R., ed.), p. 435, Reidel, Dordrecht.

Maxwell, A. and Swarup, G. (1958). *Nature*, **181**, 36.

McClements, K. G. (1987a). *Astron. Astrophys.*, **175**, 255.

McClements, K. G. (1987b). *Solar Phys.*, **109**, 355.

McClymont, A. N. and Canfield, R. C. (1983). *Astrophys. J.*, **265**, 483.

McClymont, A. N. and Canfield, R. C. (1986). *Astrophys. J.*, **305**, 936.

McIntosh, P. S. and Donnelly, R. F. (1972). *Solar Phys.*, **23**, 444.

McWhirter, R. W. P., Thonemann, P. C. and Wilson, R. (1975). *Astron. Astrophys.*, **40**, 63.

Melrose, D. B. (1980). *Plasma Astrophysics*, Gordon and Breach, New York.

Melrose, D. B. and Dulk, G. A. (1982). *Astrophys. J.*, **259**, 844.

Melozzi, M., Kundu, M. and Shevgaonkar, R. K. (1985). *Solar Phys.*, **97**, 345.

Mewe, R. (1972). *Astron. Astrophys.*, **20**, 215.

Meyer, F. (1968). In *Structure and Development of Solar Active Regions*, Proc. IAU Symp. 35 (Kiepenheuer, K. O., ed.), p. 485, Reidel, Dordrecht.

Mochnacki, S. W. and Zirin, H. (1980). *Astrophys. J. Lett.*, **239**, L27.

Montmerle, T., Koch-Miramond, L., Falgarone, E. and Grindlay, J. E. (1983). *Astrophys. J.*, **269**, 182.

Moore, R. L. and LaBonte, B. J. (1980). In *Solar and Interplanetary Dynamics*, Proc. IAU Symp. 91 (Dryer, M. and Tandberg-Hanssen, E., eds.), p. 207, Reidel, Dordrecht.

Moore, R. L. *et al.* (1980). In *Solar Flares – A Monograph from Skylab Solar Workshop II* (Sturrock, P. A., ed.), Chapter 8, p. 341, Colorado Associated University Press, Boulder, Colorado.

Moreton, G. E. (1960). *Astron. J.*, **65**, 494.

Moreton, G. E. (1964). *Astron. J.*, **69**, 145.

Morishita, H. (1985). *Tokyo Astron. Bull.*, second series, no. 272, p. 3123.

Mouradian, Z., Martres, M.-J., Soru-Escaut, I. and Gesztelyi, L. (1988). *Astron. Astrophys.*, in press.

Mouschovias, T. Ch. and Poland, A. I. (1978). *Astrophys. J.*, **220**, 675.

Munro, R. H., Gosling, J. T., Hildner, E., MacQueen, R. M., Poland, A. I. and Ross, C. L. (1979). *Solar Phys.*, **61**, 201.

Murphy, R. J. and Ramaty, R. (1984). *Adv. Space Res.*, **4**, 127.

Nagai, F. and Emslie, A. G. (1984). *Astrophys. J.*, **279**, 876.

Nakagawa, Y. and Hyder, C. L. (1969). *Envir. Res. Paper #320, AFCRL-70-0273*, Office of Aerospace Research.

Nakagawa, Y. and Raadu, M. A. (1972). *Solar Phys.*, **25**, 127.

Nakagawa, Y. and Steinolfson, R. S. (1976). *Astrophys. J.*, **207**, 296.

Nakagawa, Y., Wu, S. T. and Han, S. M. (1978). *Astrophys. J.*, **219**, 314.

Neidig, D. F. and Cliver, E. W. (1983). *Solar Phys.*, **88**, 280.

Neidig, D. F. and Wiborg, Jr., P. H. (1984). *Solar Phys.*, **92**, 217.

Neupert, W. M. (1968). *Astrophys. J. Lett.*, **153**, L59.

Obridko, V. N. (1968). *Soln. Aktivnostj*, **3** (NASA Technical Translation TT-F-581).

Ohki, K., Takakura, T., Tsuneta, S. and Nitta, N. (1983). *Solar Phys.*, **86**, 301.

Orwig, L. E. and Woodgate, B. E. (1986). In *The Lower Atmosphere of Solar Flares* (Neidig, D., ed.), p. 306, Sacramento Peak Observatory, New Mexico.

Pallavicini, R., Peres, G., Serio, S., Vaiana, G., Acton, L., Leibacher, J. and Rosner, R. (1983). *Astrophys. J.*, **270**, 270.

Pallavicini, R., Serio, S. and Vaiana, G. S. (1977). *Astrophys. J.*, **216**, 108.

Parker, E. N. (1957). *J. Geophys. Res.*, **62**, 509.

Parker, E. N. (1963). *Interplanetary Dynamical Processes*, Interscience, New York.

Parker, E. N. (1979). *Cosmical Magnetic Fields*, Clarendon Press, Oxford.

Parmar, A. N., Wolfson, C. J., Culhane, J. L., Phillips, K. J. H., Acton, L. W., Dennis, B. R. and Rapley, C. G. (1984). *Astrophys. J.*, **279**, 866.

Patty, S. R. and Hagyard, M. J. (1986). *Solar Phys.*, **103**, 111.

Perrin, F. (1942). *J. Chem. Phys.*, **10**, 415.

Peterson, L. E. and Winckler, J. R. (1959). *J. Geophys. Res.*, **64**, 697.

Petrosian, V. (1981). *Astrophys. J.*, **251**, 727.

Petschek, H. E. (1964). In *AAS-NASA Symposium on Physics of Solar Flares* (Hess, W. N., ed.), p. 425, NASA SP-50.

Pick, M. (1986). In *Solar Flares and Coronal Physics Using P/OF as a Research Tool* (Tandberg-Hanssen, E., Wilson, R. M. and Hudson, H. S., eds.), p. 159, NASA Conf. Publ. 2421.

Pneuman, G. (1980). *Solar Phys.*, **65**, 369.

Poland, A. I., Machado, M. E., Wolfson, C. J., Frost, K. J., Woodgate, B. E., Shine, R. A., Kenny, P. J., Cheng, C.-C., Tandberg-Hanssen, E. A., Bruner, E. C. and Henze, W. (1982). *Solar Phys.*, **78**, 201.

Poland, A. I., Orwig, L. E., Mariska, J. T., Nakatsuka, R. and Auer, L. H. (1984). *Astrophys. J.*, **280**, 457.

Porter, J. G., Toomre, J. and Gebbie, K. B. (1984). *Astrophys. J.*, **283**, 879.

Porter, J. G., Reichmann, E. J., Moore, R. L. and Harvey, K. L. (1986). In *Coronal and Prominence Plasmas* (Poland, A. I., ed.), p. 383, NASA Conf. Publ. 2442.

Pottasch, S. R. (1964). *Space Sci. Rev.*, **3**, 816.

Pottasch, S. R. (1965). *B.A.N.*, **18**, 8.

Potter, D. W., Lin, R. P. and Anderson, K. A. (1980). *Astrophys. J. Lett.*, **236**, L97.

Priest, E. R. and Forbes, T. (1986). *J. Geophys. Res.*, **91**, 5579.

Priest, E. R., Gaizauskas, V., Hagyard, M. J., Schmahl, E. J. and Webb, D. F. (1986). In *Energetic Phenomena on the Sun Proc. on Solar Maximum Mission Flare Workshop* (Kundu, M. and Woodgate, B., eds.), Chapter 1, NASA Conf. Publ. 2439.

Ramaty, R. (1986). In *Physics of the Sun* (Sturrock, P. A., Holzer, T. E., Mihalas, D. M. and Ulrich, R. K., eds.), Vol. 2, Chapter 14, p. 291, Reidel, Dordrecht.

Ramaty, R., Kozlovsky, B. and Lingenfelter, R. E. (1975). *Space Sci. Rev.*, **18**, 341.

Ramaty, R., Kozlovsky, B. and Lingenfelter, R. E. (1979). *Astrophys. J. Suppl.*, **40**, 487.

Raymond, J. C., Cox, D. P. and Smith, B. W. (1977). *Astrophys. J.*, **204**, 290.

Reeves, E. M., Huber, M. C. E., Timothy, J. G. and Withbroe, G. L. (1977). *Appl. Optics*, **16**, 849.

Ribes, E. (1969). *Astron. Astrophys.*, **2**, 316.

Richardson, R. S. (1951). *Astrophys. J.*, **114**, 356.

Robinson, R. (1977). *Solar Phys.*, **55**, 459.

Rompolt, B. (1984). *Adv. Space Res.*, **4**(7), 357.

Rose, W. K. (1973). *Astrophysics*, Holt, Rhinehart, and Winston, New York.

Rosenbluth, M. N., McDonald, W. M. and Judd, D. L. (1957). *Phys. Rev.*, **107**, 1.

Rosner, R., Tucker, W. H. and Vaiana, G. S. (1978). *Astrophys. J.*, **220**, 643.

Rosseland, S. (1926). *Astrophys. J.*, **63**, 342.

Rosseland, S. (1936). *Ap. Norw.*, **2**, 173.

Rust, D. M. (1973). *Solar Phys.*, **33**, 205.

Rust, D. M. (1976). *Solar Phys.*, **47**, 21.

Rust, D. M. (1982). *Science*, **216**, 939.

Rust, D., Simnett, G. M. and Smith, D. F. (1985). *Astrophys. J.*, **288**, 401.

Sahal-Bréchot, S., Bommier, V. and Leroy, J.-L. (1977). *Astron. Astrophys.*, **59**, 223.

Schadee, A. and Gaizauskas, V. (1984). *Adv. Space Res.*, **4**(7), 117.

Schmahl, E. J., Kundu, M. R., Strong, K. T., Bentley, R. D., Smith, Jr., J. B. and Krall, K. R. (1982). *Solar Phys.*, **80**, 233.

Schmidt, G. (1969). *Physics of High Temperature Plasmas*, Academic Press, New York.

Schmieder, B., Forbes, T. G., Malherbe, J. M. and Machado, M. E. (1987). *Astrophys. J.*, **317**, 956.

Schwinger, J. (1949). *Phys. Rev.*, **75**, 1912.

Severny, A. B. (1958). *Izv. Crim. Astrophys. Obs.*, **19**, 72.

Sheeley, Jr., N. R. (1966). *Astrophys. J.*, **144**, 723.

Sheeley, Jr., N. R. (1967). *Solar Phys.*, **1**, 171.

Sheeley, Jr., N. R. and Golub, L. (1979). *Solar Phys.*, **63**, 119.

Shevgaonkar, R. K. and Kundu, M. R. (1985). *Astrophys. J.*, **292**, 733.

Shoub, E. C. (1983). *Astrophys. J.*, **266**, 339.

Sime, D. G. and Hundhausen, A. J. (1987). *J. Geophys. Res.*, **92**, 1049.

Simnett, G. M. and Harrison, R. A. (1985). *Solar Phys.*, **99**, 291.

Simon, G. (1987). Ph.D. Thesis, University of Paris.

Skumanich, A. (1985). *Australian J. Phys.*, **38**, 971.

Smith, D. F. (1980). *Solar Phys.*, **66**, 135.

Smith, D. F. and Auer, L. H. (1980). *Astrophys. J.*, **238**, 1126.

Smith, D. F. and Brown, J. C. (1980). *Astrophys. J.*, **242**, 799.

Smith, D. F. and Harmony, D. W. (1982). *Astrophys. J.*, **252**, 800.

Smith, D. F. and Lilliequist, C. G. (1979). *Astrophys. J.*, **232**, 582.

Smith, D. F. and Spicer, D. S. (1979). *Solar Phys.*, **62**, 359.

Smith, H. J. and Smith, E. v. P. (1963). *Solar Flares*, MacMillan, New York.

Smith, Jr., J. B. (1986). Private communication.

Somov, B. V. (1975). *Solar Phys.*, **42**, 235.

Somov, B. V. (1981). In *Solar Max. Year*, Vol. 1, p. 155, U.S.S.R. Academy of Sciences.

Sonnerup, B. U. Ö. (1970). *J. Plasma Phys.*, **4**, 161.

Soru-Escaut, I., Martres, M.-J. and Mouradian, Z. (1985). *Astron. Astrophys.*, **145**, 19.

Soru-Escaut, I., Martres, M.-J. and Mouradian, Z. (1986). In *Solar-Terrestrial Predictions: Proceedings of a Workshop at Meudon, France, June 18–22, 1984* (Simon, P. A., Heckman, G. and Shea, M. A., eds.), p. 186, NOAA, Boulder, Colorado and Air Force Geophysics Laboratory, Hanscom AFB, Bedford, Massachusetts.

Spicer, D. S. (1976). *NRL Report 8036*, Washington, D.C.

Spicer, D. S. (1977). *Solar Phys.*, **53**, 305.

Spicer, D. S. (1982). *Space Sci. Rev.*, **35**, 349.

Spicer, D. S. and Sudan, R. (1984). *Astrophys. J.*, **280**, 440.

Spitzer, L. (1962). *Physics of Fully Ionized Gases*, 2nd edn, Interscience, New York.

Stark, J. (1916). *Jahr. Radio Elektr.*, **12**, 349.

Steinberg, J. L. and Lequeux, J. (1963). *Radio Astronomy*, McGraw-Hill, New York.

Steinolfson, R. S. and Nakagawa, Y. (1976). *Astrophys. J.*, **207**, 300.

Steinolfson, R. S., Schmahl, E. J. and Wu, S. T. (1979). *Solar Phys.*, **63**, 187.

Steinolfson, R. S., Wu, S. T., Dryer, M. and Tandberg-Hanssen, E. (1978). *Astrophys. J.*, **225**, 259.

Stenflo, J. O. (1973). *Solar Phys.*, **32**, 41.

Stepanov, V. E. (1958). *Izv. Crim. Astrophys. Obs.*, **19**, 20.

Stewart, R. T. (1972). *Proc. ASA*, **2**, 100.

Stewart, R. T. (1984). In *STIP Symposium on Solar/Interplanetary Intervals* (Shea, M. A., Smart, D. F. and McKenna-Lawlor, S. M. P., eds.), p. 253, Book Crafters, Inc., Chelsea, Michigan.

Stewart, R. T., Dulk, G. A., Sheridan, K. V., House, L. L., Wagner, W. J., Sawyer, C. and Illing, R. (1982). *Astron. Astrophys.*, **116**, 217.

Stokes, G. G. (1852). *Trans. Cambr. Phil. Soc.*, **9**, 399.

Strutt, J. (Lord Rayleigh) (1922). *Proc. Roy. Soc.*, **102**, 190.

Sturrock, P. A. (1964). In *AAS-NASA Symposium on Physics of Solar Flares* (Hess, W. N., ed.), p. 357, NASA SP-50.

Sturrock, P. (1966). *Nature*, **211**, 695.

Sturrock, P. A. (1968). In *Structure and Development of Solar Active Regions*, Proc. IAU Symp. 35 (Kiepenheuer, K. O., ed.), p. 471, Reidel, Dordrecht.

Sturrock, P. A. (ed.) (1980). *Solar Flares – A Monograph from Skylab Solar Workshop II*, Colorado Associated University Press, Boulder, Colorado.

Sturrock, P. A. and Coppi, B. (1966). *Astrophys. J.*, **143**, 3.

Sturrock, P. A., Kaufmann, P., Moore, R. L. and Smith, D. F. (1984). *Solar Phys.*, **94**, 341.

Švestka, Z. (1965). *Adv. Astron. Astrophys.*, **3**, 119.

Švestka, Z. (1966a). *Space Sci. Rev.*, **5**, 388.

Švestka, Z. (1966b). *Bull Astron. Inst. Czech.*, **17**, 262.

Švestka, Z. (1976). *Solar Flares*, Reidel, Dordrecht.

Švestka, Z. (1986). In *The Lower Atmosphere of Solar Flares* (Neidig, D. F., ed.), p. 332, Sacramento Peak Observatory, New Mexico.

Švestka, Z. and Poletto, G. (1985). *Solar Phys.*, **97**, 113.

Švestka, Z., Kopecky, M. and Blaha, M. (1961). *Bull. Astron. Inst. Czech.*, **12**, 229.

Švestka, Z., Fontenla, J. M., Machado, M. E., Martin, S. F., Neidig, D. F. and Poletto, G. (1986). *Adv. Space Res.*, **6**(6), 253.

Sweet, P. A. (1958). *IAU Symposium No. 6*, p. 123.

Symon, K. R. (1971). *Mechanics*, Addison-Wesley, Reading, Massachusetts.

Syrovatskii, S. I. (1966). *Astron. Zh.*, **43**, 340.

Syrovatskii, S. I. (1969). In *Solar Flares and Space Research* (de Jager, C. and Švestka, Z., eds.), p. 346, North-Holland Publishing Co., Amsterdam.

Takakura, T. (1969). *Solar Phys.*, **6**, 133.

Takakura, T. (1975). In *Solar Gamma-, X-, and EUV Radiation*, Proc. IAU Symp. 68 (Kane, S. R., ed.), p. 299, Reidel, Dordrecht.

Tanaka, K. (1983). In *Activity in Red-Dwarf Stars*, Proc. IAU Colloquium 71 (Byrne, P. B. and Rodono, M., eds.), p. 307, Reidel, Dordrecht.

Tanaka, K. and Nakagawa, Y. (1973). *Solar Phys.*, **33**, 187.

Tanaka, K. and Zirin, H. (1985). *Astrophys. J.*, **299**, 1036.

Tanaka, K., Watanabe, T. and Nitta, N. (1983). *Astrophys. J. Lett.*, **254**, L59.

Tandberg-Hanssen, E. (1963). *Astrophys. J.*, **137**, 26.

Tandberg-Hanssen, E. A. (1974). *Solar Prominences*, Reidel, Dordrecht.

Tandberg-Hanssen, E., Kaufmann, P., Reichmann, E. J., Teuber, D. L., Moore, R. L., Orwig, L. E. and Zirin, H. (1984). *Solar Phys.*, **90**, 41.

Tandberg-Hanssen, E. A., Martin, S. F. and Hansen, R. T. (1980). *Solar Phys.*, **65**, 357.

Tang, F. (1985). *Solar Phys.*, **102**, 131.

Tang, F. (1986). *Solar Phys.*, **105**, 399.

Tang, F. and Moore, R. L. (1982). *Solar Phys.*, **77**, 263.

Tayal, S. S. and Kingston, A. E. (1984). *J. Phys. B*, **17**, L145.

Thiessen, G. (1949). *Obs.*, **69**, 228.

Thomas, R. N. (1948a). *Astrophys. J.*, **108**, 130.

Thomas, R. N. (1948b). *Astrophys. J.*, **108**, 142.

Thomas, R. N. (1957). *Astrophys. J.*, **125**, 260.

Tindo, I. P. and Somov, B. V. (1978). *Cosmic Res.*, **16**, 555.

Tramiel, C. J., Chanan, G. A. and Novick, R. (1984). *Astrophys. J.*, **280**, 440.

Trottet, G. and MacQueen, R. M. (1980). *Solar Phys.*, **68**, 177.

Trottet, G., Avignon, Y., Kerdraon, A., Mein, N. and Pick, M. (1984). *Adv. Space Res.*, **4**(7), 271.

Trottet, G., Kerdraon, A., Benz, A. O. and Treumann, R. (1981). *Astron. Astrophys.*, **93**, 129.

Trubnikov, B. A. (1958). Ph.D. Thesis, Moscow University. (English Translation, 1960, USAEC Technical Information Service, AEC-tr-4073).

Tseng, H. K. and Pratt, R. M. (1973). *Phys. Rev.*, **A7**, 1502.

Tsuneta, S. (1983). Ph.D. Thesis, University of Tokyo.

Tsuneta, S. (1985). *Astrophys. J.*, **290**, 353.

Tsuneta, S., Takakura, T., Nitta, N., Ohki, K., Tanaka, K., Makishima, K., Murakami, T., Oda, M., Ogawara, Y. and Kondo, Y. (1984a). *Astrophys. J.*, **280**, 887.

Tsuneta, S., Nitta, N., Ohki, K., Takakura, T., Tanaka, K., Makishima, K., Murakami, T., Oda, M. and Ogawara, Y. (1984b). *Astrophys. J.*, **284**, 827.

Tsytovich, V. V. (1970). *Nonlinear Effects in Plasma*, Plenum, New York.

Uchida, Y. (1968). *Solar Phys.*, **4**, 30.

Uchida, Y. (1970). *Publ. Astron. Soc. Japan*, **22**, 341.

Uchida, Y. (1974a). In *Coronal Disturbances*, Proc. IAU Symp. 57 (Newkirk, Jr., G. A., ed.), p. 383, Reidel, Dordrecht.

Uchida, Y. (1974b). *Solar Phys.*, **39**, 431.

Uchida, Y. and Hudson, H. (1972). *Solar Phys.*, **26**, 414.

Uchida, Y., Altschuler, M. and Newkirk, Jr., G. A. (1973). *Solar Phys.*, **28**, 495.

Underwood, J. H., Antiochos, S. K., Feldman, U. and Dere, K. P. (1978). *Astrophys. J.*, **224**, 1017.

Unno, W. (1956). *Publ. Astron. Soc. Japan*, **8**, 108.

Vaiana, G. S. and Rosner, R. (1978). *Ann. Rev. Astron. Astrophys.*, **16**, 393.

van de Hulst, H. C. (1950). *B.A.N.*, **11**, 150.

van Hollebeke, M. A. I., MaSung, L. S. and McDonald, F. B. (1975). *Solar Phys.*, **41**, 189.

van Hoven, G. (1982). *Mem. S. A. It.*, **53**, 441.

van Hoven, G., Steinolfson, R. S. and Tachi, T. (1983). *Astrophys. J.*, **268**, 860.

van Hoven, G., Tachi, T. and Steinolfson, R. S. (1984). *Astrophys. J.*, **280**, 391.

van Regemorter, H. (1962). *Astrophys. J.*, **136**, 906.

Velusamy, T. and Kundu, M. R. (1982). *Astrophys. J.*, **258**, 388.

Velusamy, T., Kundu, M. R., Schmahl, E. J. and McCabe, M. (1987). *Astrophys. J.*, **319**, 984.

Vernazza, J. E., Avrett, E. H. and Loeser, R. (1976). *Astrophys. J. Suppl.*, **30**, 1.

Vernazza, J. E., Avrett, E. H. and Loeser, R. (1981). *Astrophys. J. Suppl.*, **45**, 635.

Vlahos, L. and Papadopoulos, K. (1979). *Astrophys. J.*, **233**, 717.

Wagner, W. J. (1984). *Ann. Rev. Astron. Astrophys.*, **22**, 267.

Wagner, W. J. and MacQueen, R. M. (1983). *Astron. Astrophys.*, **120**, 136.

Waldmeier, M. (1938a). *Z. Aph.*, **15**, 299.

Waldmeier, M. (1938b). *Z. J.*, **16**, 276.

Waldmeier, M. (1951). *Z. Ap.*, **28**, 208.

Walter, F., Charles, P. and Bowyer, S. (1978). *Astrophys. J. Lett.*, **225**, L119.

Warwick, J. W. (1957). *Astrophys. J.*, **125**, 811.

Warwick, J. W. and Hyder, C. L. (1965). *Astrophys. J.*, **141**, 1362.

Webb, D. F. (1985). *Solar Phys.*, **97**, 321.

Webb, D. F. and Hundhausen, A. J. (1987). *Solar Phys.*, **108**, 383.

Weiss, A. A. and Stewart, R. T. (1965). *Austr. J. Phys.*, **18**, 143.

Weisskopf, V. F. I. and Wigner, E. (1930). *Z. Phys.*, **63**, 54.

White, O. R. (1961). *Astrophys. J.*, **134**, 85.

Wild, J. P. (1969). *Proc. Astron. Soc. Austr.*, **1**, 181.

Wild, J. P. and McCready, L. L. (1950). *Austr. J. Sci. Res.*, **A3**, 387.

Wild, J. P., Sheridan, K. V. and Neylan, A. A. (1959). *Austr. J. Phys.*, **12**, 369.

Wild, J. P., Smerd, S. F. and Weiss, A. A. (1963). *Ann. Rev. Astron. Astrophys.*, **1**, 291.

Williams, A. C. (1986). *IEEE Trans. Plas. Sci.*, **PS-14**(6), 838.

Wilson, R. M. (1987). *Planet. Space Sci.*, **35**, 329.

Wilson, R. M. and Hildner, E. (1984). *Solar Phys.*, **91**, 169.

Wilson, R. M. and Hildner, E. (1986). *J. Geophys. Res.*, **91**, 5867.

Withbroe, G. L. (1970). *Solar Phys.*, **11**, 42.

Withbroe, G. L. (1975). *Solar Phys.*, **45**, 144.

Withbroe, G. L. (1977). In *Proceedings of the OSO-8, November 7–10, 1977 Workshop* (Hansen, E. and Schaffner, S., eds.), p. 2, Laboratory for Atmospheric and Space Physics, Boulder, Colorado.

Withbroe, G. L. (1978). *Astrophys. J.*, **225**, 641.

Withbroe, G. L., Habbal, S. R. and Ronan, R. (1985). *Solar Phys.*, **95**, 297.

Wolfson, C. J., Doyle, J. G., Leibacher, J. W. and Phillips, K. J. H. (1983). *Astrophys. J.*, **269**, 319.

Wu, S. T., Dryer, M., Nakagawa, Y. and Han, S. M. (1978). *Astrophys. J.*, **219**, 324.

Wu, S. T., Nakagawa, Y., Han, S. M. and Dryer, M. (1982). *Astrophys. J.*, **262**, 369.

Yeh, T. and Dryer, M. (1981). *Astrophys. J.*, **245**, 704.

Zarro, D. M. and Zirin, H. (1985). *Astron. Astrophys.*, **148**, 240.

Zheleznyakov, V. V. (1970). *Radio Emission of the Sun and Planets*, Pergamon Press, New York.

Zirin, H. and Ferland, G. J. (1980). *Big Bear Solar Observatory Preprint No. 0192*, Pasadena, California.

Zirin, H. and Tanaka, K. (1973). *Solar Phys.*, **32**, 173.

Zirin, H. and Tandberg-Hanssen, E. (1960). *Astrophys. J.*, **131**, 717.

Zvereva, A. M. and Severny, A. B. (1970). *Izv. Crim. Astrophys. Obs.*, **41**, 97.

Author index

Subject index